Effective Procurement of Ground Investigation

emerald
PUBLISHING

ice
Publishing

Effective Procurement of Ground Investigation

Prepared by
The Association of Geotechnical and
Geoenvironmental Specialists

Published by Emerald Publishing Limited, Floor 5,
Northspring, 21–23 Wellington Street, Leeds LS1 4DL.

ICE Publishing is an imprint of Emerald Publishing Limited

Other ICE Publishing titles:
UK Specification of Ground Investigation, third edition.
The Association of Geotechnical and Geoenvironmental
Specialists. ISBN 978-0-7277-6523-9
*Safe Intrusive Activities on Land Potentially Impacted by
Contamination.*
British Drilling Association. ISBN 978-0-7277-3507-2.
ICE Manual of Geotechnical Engineering, second edition.
Edited by Michael Brown, John Burland, Tim Chapman, Kelvin
Higgins, Hilary Skinner and David Toll. ISBN 978-0-7277-6685-4

A catalogue record for this book is available from the British Library

ISBN 978-1-83662-029-7

Cover photo: Cover image of rotary drilling, courtesy of Chris
Wade, Soil Engineering Geoservices Limited.

Commissioning Editor: Michael Fenton
Content Development Editor: Cathy Sellars
Production Editor: Emma Sudderick

Typeset by KnowledgeWorks Global Ltd.
Index created by David Gaskell

Contents

Foreword

This document is based on the Site Investigation Steering Group (SISG) publication '*Effective Site Investigation*' which formed part of the Site Investigation in Construction series, republished in 2016 by ICE Publishing. Since much of the original document has been superseded by more recent industry publications, this document represents a major revision and includes the aim of bringing together good industry practice with regard to improving procurement of ground investigations.

The benefit of good ground investigation for the design of structures, environmental solutions, assessment of land quality and ground stability is well established. Delivery of good ground investigations is often hindered by the procurement process where cost over quality is chosen and when the process is led by personnel with a poor understanding of how the benefits are realised. Whether the project requires an assessment of ground conditions for geotechnical or land contamination, the process should be driven by an overriding requirement to assess and reduce uncertainties and the delivery of high-quality reliable data.

This document is intended primarily for Clients but construction professionals such as civil and structural engineers, architects, developers and other procurers will benefit from using this guidance. Ground engineering professionals will also find the document useful but will need to reference more detailed information elsewhere – for example, *UK Specification for Ground Investigation*.

Those involved in the procurement process must recognise that all ground investigations require combinations of equipment, methods and specialists – for example, ground investigations specifically for land contamination or waste assessments would require involvement of geoenvironmental specialists, sampling would demand different techniques, transport and storage requirements, and analytical laboratory testing for contaminants would require much shorter timescales.

The procurement of successful ground investigations can only be accomplished when there is collaboration between all stakeholders, an understanding of the complete ground investigation process, appreciation of realistic timescales, careful planning, and the employment of properly equipped and competent personnel at all stages. It is essential that competent ground practitioners with proven experience, relevant to the ground investigation required, are involved as early as possible and throughout, from planning through to post construction. Careful consideration should also be given to the conditions of contract adopted and the contractual relationships so that good communication and quick decision making are not inadvertently hindered to the detriment of the desired outcomes.

This document adopts the roles identified within the Construction (Design and Management) (CDM) Regulations 2015. The Client shall be the procurer of the ground investigation and shall appoint in writing specialists to develop, execute and manage the ground investigation. The Client shall adopt the role of, or appoint, a Designer (they may also be appointed as the Principal Designer) to develop the scope of the ground investigation. A Contractor shall also be appointed (they may also be appointed as the Principal Contractor) to execute and manage the ground investigation. The Client shall also appoint a ground investigation adviser who may be part of their team or provided from a third party and may act as the Investigation Supervisor. It has been assumed that technical direction of the investigation will be the responsibility of the Investigation Supervisor: an experienced ground practitioner who may be someone appointed by the Client, Designer, Contractor or an independent organisation. The Investigation Supervisor is required to independently assist and advise the Client and Designer. It should be recognised that organisations and individuals may also be given additional role names, such as contract-related roles (e.g. the 'Engineer') and, therefore, reference to the specific conditions of contract should be made.

This publication is intended to assist with the procurement of effective ground investigations which identify and minimise commercial, health, safety and environmental risk as well as providing sufficient technical data to enable efficient design and/or assessment of natural ground and groundwater hazards as well as land potentially impacted by contamination.

Acknowledgments

This document was funded and written by the Association of Geotechnical and Geoenvironmental Specialists (AGS) in collaboration with the British Drilling Association and British Geotechnical Association.

This edition is based on the Site Investigation Steering Group publication 'Effective Site Investigation' but has been significantly revised. This revision was carried out by the AGS Revision Working Group, part of the AGS Procurement of Ground Investigation Steering Group. The Revision Working Group comprised:

Lead Authors:

Name	Affiliation & Company
Julian Lovell	Equipe Group & Association of Geotechnical and Geoenvironmental Specialists
Matthew Baldwin	Independent Consultant & Association of Geotechnical and Geoenvironmental Specialists

Working Group Leaders:

Name	Affiliation & Company
Jamie Codd	AECOM
Adam Latimer	Ian Farmer Associates & Association of Geotechnical and Geoenvironmental Specialists
Neil Parry	Geotechnical Engineering & Association of Geotechnical and Geoenvironmental Specialists
Phil Smith	Geotechnical Consulting Group & British Geotechnical Association

Working Group Members:

Name	Affiliation & Company
Simon Baxter	Soil Engineering
Natalie Bews	Concept Consultants
Paul Bradley	Jacobs
Jon Christie	BAM
Patrick Cox	AECOM

Name	Affiliation & Company
Anthony Drake	Mott MacDonald
Neil Esslemont	East West Rail
Alice Hayton	Arcadis
Angus Higgins	National Highways
Andy Indoe	WSP
Stewart Jarvis	Arup
Joseph Martin	Network Rail
Catherine McGrath	National Highways
Paul McMann	Dunelm & British Drilling Association
Mary Nicholls	WSP
Richard Nicholson	AECOM
Richard Patterson	Mott MacDonald
Tracey Radford	Atkins Realis
Peter Reading	PJCR Consulting & SOCOTEC
Mark Postle-Hacon	Structural Soils
Peter Rhodes	Raeburn Drilling & Geotechnical
Ian Samworth	Network Rail
Hilary Shields	Tony Gee and Partners
Gugulethu Sibanda	SOCOTEC
Henry Tayler	Arup
Mark Toye	SOCOTEC & British Drilling Association
Michael Vance	Mott MacDonald
Rob Williams	Structural Soils
Wesley Wray	HS2

The authors acknowledge the contributions made by many other colleagues of the AGS Revision Working Group members, the AGS Safety Working Group and the AGS Sustainability Working Group during the drafting of the document.

All images courtesy of © AGS Member Photograph Competitions and individual provider acknowledgements.

Association of Geotechnical and Geoenvironmental Specialists
ISBN 978-1-83662-029-7
https://doi.org/10.1108/978-1-83662-026-620251001

Chapter 1
Introduction

The importance of ground investigation in the construction process is now well established. Numerous projects have well-documented details of cost overruns due to unexpected/unforeseen ground conditions and numerous publications including ACE (2011), Alhalaby and Whyte (1994), Ashton and Gidado (2001), Clayton (2001), John *et al.* (2024), Kinlan (2014), Maliphant *et al.* (2015), Mott MacDonald and Soil Mechanics (1994), Peacock and Whyte (1992) and Tyrrell *et al.* (1983) have provided evidence to this effect. John *et al.* (2024) has also, more recently, discussed the link between well-designed investigations and significant reductions in project outturn costs. This link is shown in Figure 1.1.

Many industry practitioners suggest that the typical cost of a ground investigation is in the range of 1–2% of the project cost, but that project cost increase due to ground related issues is typically in the range of 10–50%. The research and data obtained shows that the cost of ground and groundwater-related issues often significantly and disproportionately impacts on project outturn costs but when the cost of the ground investigation increases this risk is significantly reduced. However, the data also indicates that the typical cost for a ground investigation is actually between 0.11 and 0.33% of the project cost, which indicates that Clients are still not recognising the benefits of procuring adequate ground investigations.

It is not just project costs, however, that are affected by the adequacy of the ground investigation. Appropriate ground investigation in relation to any specific project will reduce risk (see Chapter 3), enhance health and safety for the construction workforce, increase quality, protect the environment, help to meet sustainability goals, as well as enhancing public perception of the enterprise.

Figure 1.1 Relationship between ground investigation cost as a percentage of project award cost and project escalation costs (Mott MacDonald and Soil Mechanics, 1994)

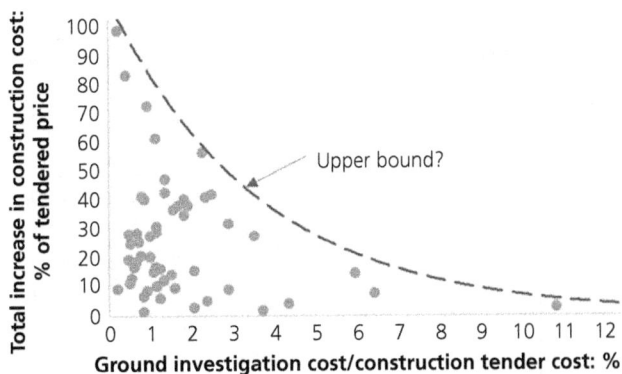

Although the importance and relevance of good ground investigation is becoming embedded in the 'thought processes' of many organisations involved in the construction process, the method of procuring such investigations still leaves much room for improvement. It is vital that the procurement process involves not only the relevant stakeholders, but also that these stakeholders create a team that is competent to specify and procure ground investigations that are relevant to the projects in question. Table 1.1 shows the types and sources of geotechnical risk. It is worth noting that all the issues marked with (*) result from failings with one or more of the stakeholders in the ground investigation procurement process.

Table 1.1 Sources of geotechnical risk (after Baynes, 2010)

Type of geotechnical risk		Hazard	Source
Project management		Poor management of entire geo-engineering process	An inadequate understanding of the importance of ground conditions (*), resulting in poor management of the entire geo-engineering process – for example, a decision to submit a tender price with no risk weighting for geotechnical factors
Contractual		Poor management of ground investigation and contract documentation	An inadequate understanding of the importance of ground conditions (*), resulting in poor acquisition, understanding and/or communication of ground investigation information; this often leads to claims based on unforeseen ground conditions
Technical	**Analytical**	Unreasonable analytical model chosen	An inadequate understanding of ground conditions (*) and analytical methods, resulting in an unreasonable choice of analytical models
	Properties	Unreasonable design values chosen	An inadequate understanding of ground conditions (*) and field and laboratory testing, resulting in an unreasonable choice of design values
	Geological	Unforeseeable geological details	Geological conditions are variable and investigation of all geological details is impracticable
		Inherently hazardous ground conditions	Geological conditions and geological processes that involve hazards such as large ground movements, voids, aggressive chemistry, erosion and so on
		Unforeseen ground conditions	Inadequate understanding of geological conditions (*), resulting in unforeseen ground conditions being encountered during construction, often because of an inadequate ground investigation due to poor management

Professionals involved with the construction process typically comprise architects, structural and civil engineers, and building contractors, all of whom ultimately work for the Client. All too often, ground engineering practitioners are missed out from this list when investigations are procured. When this happens and investigations are procured by individuals who lack a complete understanding of appropriate ground investigation techniques in relation to ground conditions and/or the proposed structures, inadequate investigations ensue.

With so much historical and recent evidence linking the value of properly procured ground investigations to enhanced construction key performance indicators, this document is intended to provide a valuable tool in the overall procurement process. The contents have been carefully considered to provide up-to-date guidance on the procurement process. By drawing on examples of where best practice in procurement has been adopted, this document will aid the user by highlighting the key considerations when embarking on the ground investigation procurement process.

There are various standard conditions of contract that may be used to procure a ground investigation, each with their own advantages and disadvantages. It is helpful if the conditions of contract selected are readily able to cope with the unknowns and inherent variability in scope of a ground investigation.

Chapter 5 provides guidance on typical contracts used, together with their advantages and limitations when used for ground investigations.

This publication references and signposts other key industry publications to assist the user in developing ground investigations that are 'fit for purpose' and provides an industry best practice guide to the effective procurement of ground investigations. It is intended for use by Clients and all those construction professionals that either procure or have a stake and an interest in ensuring that ground investigations deliver information key to the design process and risk management process.

REFERENCES

ACE (Association for Consultancy and Engineering) (2011) *Infrastructure Conditions of Contract: Ground Investigation Version.* ACE, London, UK.

Alhalaby NMH and Whyte IL (1994) The impact of ground risks in construction on project finance. In *Risk and Reliability in Ground Engineering*: Proceedings of the conference organized by the Institution of Civil Engineers, held in London 11–12 November 1993. Thomas Telford, London, UK, pp. 54–67.

Ashton P and Gidado K (2001) Risk associated with inadequate site investigation procedures under design and build procurement systems. *Proceedings of Association of Researchers in Construction Management Conference, Salford, UK.*

Baynes FJ (2010) Sources of geotechnical risk. *Quarterly Journal of Engineering Geology and Hydrogeology* **43(3)**: 321–331.

Clayton CRI (2001) *Discussion: Managing Geotechnical Risk: Time for Change?* Institution of Civil Engineers (ICE), London, UK.

John B, Tim C, Hilary S and Michael B (2024) *ICE Manual of Geotechnical Engineering*, 2nd edn. Emerald Publishing/ICE Publishing, Leeds, UK, vol. 1.

Kinlan D (2014) *Adverse Physical Conditions and the Experienced Contractor*. Delft Academic Press, Nieuwerkerk aan den Ijssel, the Netherlands.

Maliphant PC, Eynon M, Hannam D and Wu E (2015) Managing geotechnical risk: Contemporary challenges. In *ICE Manual of Geotechnical Engineering*. Institution of Civil Engineers (ICE), London, UK.

Mott MacDonald and Soil Mechanics (1994) *TRRL Project Report 60. Study of the Efficiency of Site Investigation Practices*. Transport Research Laboratory, Crowthorne, UK.

Peacock WS and Whyte IL, CASPAR (1992 May) Site investigation and risk analysis. *Proceedings of the Institution of Civil Engineers - Civil Engineering* **92(2)**: 74–81.

Tyrrell AP, Lake LM and Parsons AW (1983) *An Investigation of the Extra Costs Arising on Highway Contracts*. Transport Road and Research Laboratory, Crowthorne, UK, TRRL Supplementary Report SR814.

Association of Geotechnical and Geoenvironmental Specialists
ISBN 978-1-83662-029-7
https://doi.org/10.1108/978-1-83662-026-620251002
Emerald Publishing Limited: All rights reserved

Chapter 2
Planning

2.1.   Introduction

Planning is key to the successful procurement and execution of any ground investigation. The thought process that identifies exactly what is required in terms of input and output from a ground investigation should be transparent to all parties involved and must be documented so that what has been considered is available for reference.

The input to the planning of the ground investigation, in terms of project team, is likely to depend on the size and complexity of the project. Factors affecting the complexity of a project may include

- the nature and variability of the specific ground conditions anticipated
- the importance and sensitivity of the project
- the size of the project
- the complexity of the working arrangements (e.g. working airside at an airport, over water, etc.).

A ground investigation for a large linear infrastructure project is likely to require a significantly greater level of expertise in the planning than a ground investigation for a small housing project. Similarly, the ground investigation for a project involving complex analysis and complex ground conditions might be expected to require a greater level of specialism than is required for a project involving simple ground conditions or analysis. A structured planning process should be followed, regardless of project size and complexity.

2.2.   CTQ (cost, time and quality) triangle
2.2.1   Introduction

The Project Management Triangle developed by Barnes (1969) has been modified in Figure 2.1 to reflect procurement within ground investigation. It shows three elements of procurement – cost, time and quality (CTQ) – each of which when altered will affect the scope. Balancing the so-called CTQ triangle so that all stakeholders are satisfied is the holy grail for most projects. If this process is carried out properly during planning, the output document (which is the scope of the ground investigation) should provide a framework for delivering a ground investigation that is wholly appropriate to the site in question. By following a structured planning methodology regarding procurement (as described within this section), the key elements of cost, time and quality can be achieved for all projects.

2.2.2   What goes wrong?

Very often, one or more of the elements of the CTQ triangle get skewed – for example, the time available for the investigation may get shortened, and this can have an inverse effect on the cost – that is, the cost increases as more resources are required. At the same time, overall quality of the investigation may be reduced as less time is available to perform tasks correctly.

There is, however, another variable that is often not included in the CTQ triangle, namely risk, as shown in Figure 2.1. Every facet of ground investigation procurement is affected to a greater or lesser extent by the perceived risk and this can be seen as a 'cloak' that wraps around all the CTQ elements.

Figure 2.1 Modified cost, time and quality (CTQ) triangle for ground investigations

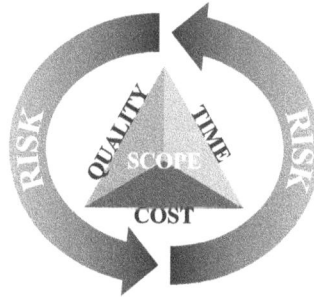

Alterations to the cost, time or quality elements of the procurement process can also change the level of risk. In many ways risk is even more emotive and important to Clients than cost, since perceived risk often shapes the whole thought process for a project. In extreme cases, the perception that risk is too great and will impact on long-term trading can lead to the abandonment of a ground investigation.

2.2.3   The ideal

Ideally, when cost, time and quality are being considered for a new project, in-depth risk assessment should be added in at the same time. By considering and including risk in the process, a more complete procurement process is achieved and this will inevitably lead to a more robust ground investigation document for all stakeholders to use. Chapter 3 provides guidance to help Clients identify and manage risk.

2.3.   Ground investigation adviser

It is essential that the Client has a clear understanding of why the ground investigation is required and what its objectives are for them to procure an effective ground investigation. To achieve this, the Client should employ or consult with a ground investigation adviser. This individual or organisation, often the Designer, should be an experienced ground practitioner who is formally appointed to ensure that ground-related risks are fully understood and addressed by the proposed scope of investigation. It is beneficial for the ground investigation adviser to communicate the objectives to the other stakeholders as well as the Client so that they understand what they are paying for and why, and to the ground investigation contractor so that they can ensure that the objectives are achievable. They should communicate not only what is required, but also why it is required. This is particularly important on large/complex projects where there may be multiple stakeholders each seeking information on ground conditions. In such circumstances, good communication and coordination can generate efficiencies in the delivery of the ground investigation, perhaps the most obvious example being ensuring that geotechnical and geoenvironmental issues are addressed by a single combined investigation where feasible, rather than as separate investigations. The ground investigation adviser may on some projects also act as the Investigation Supervisor.

2.4. Managing ground investigations

In developing the scope of a ground investigation, Clients should ensure that planning considers arrangements that are appropriate to manage both the safety and quality of the work. The requirements may come from several of the project stakeholders, including

- Client requirements
- Designer's requirements
- Contractor's processes
- third party requirements (e.g. Network Rail, Transport for London, Coal Authority, utility companies, landowner, etc.).

The relative impact of the different stakeholders will vary from project to project and may depend on the perceived level of risk that a project presents to each stakeholder. It might be expected that more onerous requirements might be put in place where the stakeholder perceives the risk to be greater. This risk might be technical, safety related or risk to the stakeholder's operations (Burland et al., 2024).

2.5. Creating the competent team

There are a number of roles required for the delivery of an effective ground investigation; some are defined in legislation – for example, Construction (Design and Management) Regulations, 2015 (CDM Regulations) (HSE, 2015a), Control of Asbestos Regulations, 2012 (HSE, 2012); others are set out in the specification used – for example, UK Specification for Ground Investigation (AGS, 2022) (*UK Specification*) or the Conditions of Contract. The Client has a legal obligation as well as a commercial interest to ensure that these roles are carried out by competent individuals. Competence is a combination of experience, skills, ability, training and qualifications (professional or vocational).

The CDM Regulations define roles for the safe delivery of a construction project (including ground investigation). It is important to note that anyone appointed to undertake roles within the scope of the regulations is required by law to have the knowledge, skills, experience and organisational capability to fulfil the requirements of the role.

The main roles defined in the CDM Regulations are

- Client
- Principal Designer (if required)
- Designer
- Principal Contractor (if required)
- Contractor.

A description of the roles and their requirements is beyond the scope of this publication. For more information see *Managing Health and Safety in Construction: Construction (Design and Management) Regulations 2015: Guidance on Regulations L153* (HSE, 2015b).

The *UK Specification* defines roles for the delivery of a ground investigation project. A key role is that of an Investigation Supervisor who is defined as

the named individual having responsibility to see that the technical objectives and quality of the investigation are met with the programme and cost constraints.

7

The *UK Specification* goes on to state that the Investigation Supervisor requires

> an appropriate level of knowledge and working experience... [and] they may require the assistance of one or more specialists... dependent on the nature, size and complexity of the investigation.

The *UK Specification* makes provision for several other roles who may be engaged in the project including a number of supervisory roles, such as

- project manager
- planner/programmer
- site agent
- health, safety and environment coordinator
- technical lead
- drilling supervisor
- technical drilling supervisor
- logging and sampling supervisor
- field testing supervisor
- instrumentation and monitoring supervisor
- other specialists.

The requirement for these roles should be determined as part of the process for scoping the ground investigation to address identified risks or areas that may be particularly critical to the interpretation of the ground investigation results – for example, a logging and sampling supervisor might be required where there are concerns about ensuring consistency of logging or the need to recover high quality samples for specialist laboratory testing; a field testing supervisor might be required where the scope includes less common field (in situ) testing, or where the results of the field tests are particularly critical to the design.

However, the ground investigation contractor might decide to make use of such specialists as part of their own assurance or quality management processes or to develop the competence of their staff through training and mentoring.

As with the appointment of the Investigation Supervisor, the use of additional supervisory staff should be dependent on the nature, size and complexity of the works being carried out.

When assessing the competence of the Investigation Supervisor, Clients should check their professional qualifications and experience. Clients can further check if they are registered with the UK Register of Ground Engineering Professionals (RoGEP) (ICE, 2023) or, for environmental investigations, a Specialist in Land Condition (SiLC, 2024) as a means of demonstrating their competences.

Larger, more complex or more critical projects may necessitate higher levels of competence for those undertaking the works. In addition, for larger, more complex or more critical projects, it may be appropriate to follow a formal competence assessment process to manage the competence of those undertaking the work. Even for the simplest projects, it is essential that all those involved have a sufficient level of experience to allow them to carry out their duties safely and competently.

It should also be expected that supervision should be provided by staff who have the appropriate level of competence.

The *UK Specification* makes provision for specifying minimum requirements for the staff required for project management, supervision and execution of the project. Those specifying a ground investigation should consider specifying minimum levels of experience required for the staff carrying out the work. This should include as a minimum

- specifying that logging staff should have attended an appropriate formal training course in logging and can demonstrate previous experience of logging the particular materials that are likely to be encountered in the project
- specifying that operators of drilling machines and drilling supervisors hold a relevant vocational qualification in land drilling
- specifying that operators of drilling machines hold a current audit card issued by the British Drilling Association
- specifying that geotechnicians including field, laboratory, cone penetration testing, construction materials and instrumentation and monitoring technicians hold the relevant vocational qualification
- specifying that supervisors can demonstrate previous experience and hold a relevant qualification
- specifying a minimum period of relevant postgraduate experience for staff, including management, supervision, testing and logging
- specifying a quality assurance programme – for example, check logging, sample quality, test results and so on by a suitably experienced person.

For larger, more complex or critical projects, such as long linear infrastructure projects, tunnels, deep excavations or highly contaminated sites, the Client should consider additional measures to ensure the quality and consistency across the project, such as

- project-specific logging training for staff: this should include staff working for the ground investigation contractor, the Designer, the Investigation Supervisor and sometimes the Client
- procuring relevant specialists to provide training to drilling, logging or testing staff on specific requirements or ground conditions or strata that will be encountered; these specialists might have a background in academia, industry or a national geological survey
- pre-use inspection of plant by an independent body to confirm compliance with legal requirements, particularly Provision and Use of Work Equipment Regulations 1998 (PUWER) (HSE, 1998) and confirmation that equipment is calibrated.

For large, geographically dispersed projects or where investigations are undertaken over multiple phases, it is possible that multiple Contractors will undertake the ground investigations. Project-specific training and the retention of strata and formation specialists to advise and check-log can be invaluable in such circumstances to ensure consistency of the deliverable from each Contractor.

2.6. Importance of supervision

The role of supervision for a ground investigation should not be underestimated by Clients and should always be provided for all types and sizes of projects. Chapter 6, Section 6.5 provides further guidance on the importance of adequate supervision.

Contractors may have a commercial incentive to provide supervision to meet their contractual obligations and most definitely have a legal obligation to provide supervision if they are employing inexperienced or trainee staff for parts of the work. However, often the financial constraints result in an inadequate level of supervision unless there is a mechanism to receive payment for these staff, and where additional supervision would enhance or ensure the success of the ground investigation from both a quality and safety perspective. The planning process should assess the level of supervision required for an effective ground investigation and which parties are responsible for covering the supervision costs.

2.7. Scoping the ground investigation

The scoping of the ground investigation should be developed from the purpose and objectives defined and agreed with the Client. Making use of standard documents such as the *UK Specification* and standard conditions of contract should help this process. Chapter 5 of this publication provides standard clauses in relation to common ground investigation activities in contracts, reducing the need for specific clauses to be developed for each project. As part of this process, all relevant project roles should be identified. Chapter 6 provides details on developing the actual specification once the scope has been determined.

Table 2.1 provides an overview of the advantages and limitations of common ground investigation techniques. It is important that Clients appreciate that the type of technique will impact on the cost, quality of the data and duration of the investigation. Clients should also appreciate that this is only one part of the process and that all other aspects need to be considered during planning, as detailed in the following sections.

2.8. Programme

Ensuring clarity and unanimous acceptance of programme requirements and key delivery dates is paramount for all parties involved. Deviations from the programme are often a source of contention but often the discontent or contention is a direct result of poor communication. Programmes should not be static documents generated at the project's inception and then left untouched. Instead, they should be dynamic, living documents requiring continuous updates to reflect actual durations, accounting for delays, changes in scope and variations.

In the context of ground investigation programmes, it is essential not only to outline the Contractor's key deliverables and their respective durations but also to encompass external factors influencing project delivery. This includes factors like the time required for document reviews, site access dates, dates for external licenses/permissions and delivery of instructions (i.e. laboratory schedules).

Many projects have constraints on the programme that may require appropriate time and resources to manage, including

- land access availability
- time to collate preconstruction information
- time to make appropriate safety arrangements, including welfare facilities

- supplier availability
- pre-works environmental or archaeological surveys
- resources required, including activities that cannot run concurrently
- time required for laboratory testing
- time required for monitoring structures or ground
- when are the results from the investigation required?
- should the investigation be delivered in several phases or are there sequencing requirements?

Where there is a large amount of work to be completed in a relatively short time frame, such that multiple resources are required to complete the work, significant effort may be required to coordinate those resources efficiently and safely.

2.9. Compliance with legal requirements

There are a wide range of legal requirements affecting ground investigation projects and it is not possible to list all of them here. The requirements derive from a variety of sources including

- health and safety legislation (e.g. Health and Safety at Work etc. Act 1974 and associated regulations, including Construction (Design and Management) Regulations 2015)
- legislation associated with the site usage (e.g. rail, mines and quarries, nuclear, military)
- environmental legislation.

Some of the legislative requirements may require specific competent staff to comply with the requirements of the legislation, such as safety-critical staff in the rail environment. On other projects, it may be desirable to involve specialists from appropriate disciplines to manage compliance with the legislation – for example, an ecologist to identify protected species or habitats.

2.10. Reporting and data

2.10.1 Reporting

The end product of any ground investigation is invariably some form of report. In seeking to achieve effective procurement of the investigation, it is, therefore, important to have a clear idea before the investigation commences of how the investigation is to be reported and what information is to be included.

However, it is also important to understand how the information from the ground investigation is used; the reporting of the ground investigation that is specified before the investigation commences should be suitable and sufficient for the intended use – for example, a project that involves a Eurocode-compliant design should have a ground investigation report (GIR) compliant with Eurocode 7 (BSI, 2024), which in turn forms part of the geotechnical design report (GDR). However, not all projects that require a ground investigation are completed to full Eurocode 7 design requirements, and the traditional 'factual report' remains a common required output, particularly for smaller projects and geoenvironmental ground investigations.

The specifier of the ground investigation should understand how the results of the ground investigation are to be used, since if the reporting is inappropriate to the design requirement, there can be cost and delay in resolving inconsistencies between what was actually reported compared to what was needed.

Table 2.1   Overview of the advantages and limitations of common ground investigation techniques

Ground investigation method	Depth: m[1]	Geotechnical data	Geoenvironmental data	Soils	Rock/hard strata	High-quality samples[2]	High-volume samples[3]	Strata identification	High-quality field testing[4]	Groundwater data	Instrumentation	Speed of progress	Sensitive sites[5]	Restricted sites[6]	Environmental impact[7]	Perceived cost[8]
Hand excavated trial pits	<3	●	●	●		●	●	●		o		M	●	●	L	M
Machine excavated trial pits	<5	●	●	●		●	●	●		o		H		o	H	L
Hand auger boreholes	<5	●	●	●							●	H	●	●	L	L
Dynamic probe borehole	<15	●		●					o			H	●	●	L	L
Dynamic sampling boreholes	<10	●	●	●		o		●	o	●	●	H	●	●	L	L
Cable percussion boreholes	<30	●	●	●		●	o	●		●	●	M	o	o	M	M
Rotary boreholes – non cored	>30	●	●	●	●			●	o	●	●	H	o	o	M	M
Rotary boreholes – cored	>30	●	●	●	●	●		●	●	o	●	M	o	o	M	H
Sonic boreholes	>30	●	o	●	●	o		●	●	●	●	H	o		L	H
Cone penetration test (CPT)	>30	●	●	●		o		●	●	●	●	H	o	o	L	H
Surface geophysics	<3	●	o	●	●			o	●			H	●	●	L	H

● denotes method is suitable

o denotes method may be suitable but will require additional equipment or specialist machine

L = Low; M = Moderate; H = High

1. Depth will depend on the ground conditions (including groundwater), actual machine used and competency of the operator.
2. High-quality samples are required to carry out laboratory strength, stiffness and compressibility tests.
3. High-volume samples would typically be in excess of 25 kg.
4. High-quality field testing includes pressuremeter, dilatometer, penetration vane, CPT and geophysics.
5. Sensitive sites include SSSIs, heritage sites etc.
6. Restricted sites include narrow access, slopes and confined spaces.
7. Environmental impact includes damage to surface conditions, potential pollution and difficulty of reinstatement.
8. Cost must be considered against the requirement to meet the ground investigation objectives.

2.10.2   Digital data

It is important to recognise that the true output of a ground investigation is not one or more reports, but data. In procuring a ground investigation, it is critical that the procuring parties understand what data is required, who will use the data and how. As shown in Figure 2.2, this aspect links many of the stages or phases involved in a ground investigation and issues previously discussed: knowing who will use the data requires an understanding of the ground investigation teams and the wider project team. Knowing what the data is to be used for and how is likely to directly impact on methodology of investigation, and certainly affects the reporting requirements.

The standard format for the electronic transfer of most ground investigation data is the Association of Geotechnical and Geoenvironmental Specialists (AGS) data format. However, the Client should consider who will be responsible for checking that data and who will maintain the project database. Where the AGS data format is not available for certain data, the Client should ensure that it is specified how that data should be transferred. On larger projects, multiple phases of investigation can occur, potentially with multiple Contractors involved. There should be a single, competent, controlling party ensuring that all the data from the disparate sources is merged into a single, secure database, and that access to this data is provided to all parties that need it, in a timely and efficient fashion. This is a distinct work activity that should be allowed for commercially, with an appropriate party appointed to undertake the management, and time/funds allowed for it.

For geoarchaeological projects there is currently no single format for data, although commercial archaeological companies typically use formats including AGS that are compatible with a range of common database and modelling software. Archiving is a requirement of planning policy (NPPF, 2023, Section 16, 211) and the Archaeology Data Service (ADS, 2024) provides a secure

Figure 2.2 The ground investigation geowheel

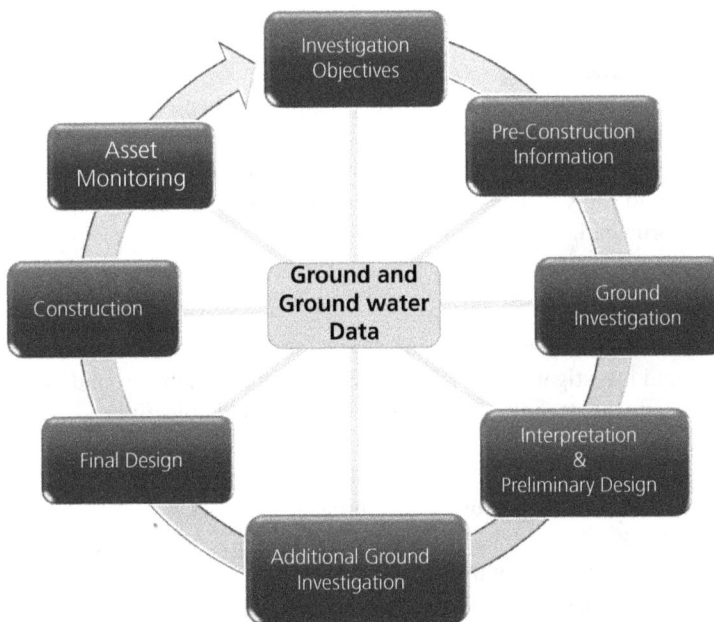

digital archiving facility (Historic England, 2020). The key elements of geoarchaeological ground investigation reporting are a clear statement of potential and significance that can inform decision making and a commitment to archive data.

2.10.3   Data sharing

While most ground investigations are stand-alone activities, in some cases, a project may be being developed in close proximity to another. This can be particularly the case with larger projects, possibly involving large infrastructure or other publicly procured projects. In such cases, there is a risk that, if both projects operate in isolation to each other, important information about ground conditions identified by one project may not become known to the other. On a more basic level, savings in the cost of the ground investigations may be possible if the projects communicate and agree a data-swap, with less duplication of investigation works.

Thus, where feasible, it is also advised to contact any other projects engaged in ground investigations to determine the potential for data-swaps.

A specific element of this is that the British Geological Survey maintain a database of AGS data which can be accessed (typically at the desk study stage). Such data is clearly of great use to anyone where there is data close to their project site, but its value is dependent on the extent of the database. The Construction Playbook (2022) requires that all data from investigations procured for public clients must be submitted to the UK digital data portal (http://transfer.bgs.ac.uk/ingestion) of the National Geoscience Data Centre (NGDC) of the British Geological Survey (BGS), to become part of the UK repository of subsurface data for re-access and longevity. To avoid possible disputes over data ownership and intellectual property rights, it should be stated as a contractual requirement set down by the Client that all AGS data is so transferred. The contractual requirement should clearly state who is responsible for providing the data to the British Geological Survey. Where there is a specific data controller for the project, it is likely to be most sensible for them to transfer the data to the British Geological Survey on completion of all investigation works; for smaller projects, it might be appropriate to require the Contractor to directly transfer the data.

2.11.   Effective planning

Effective planning for a successful and effective ground investigation should follow ten key steps in the procurement process which are summarised below. Figure 2.3 expands on these ten steps and provides detailed consideration which the Client should address during this process. It also provides signposting to the Chapters within this document where these steps are discussed and further guidance is provided.

The ten key steps are as follows.

1. Identify the general ground-related requirements for the project or structure.
2. Appoint a ground investigation adviser to design and manage the ground investigation.
3. Identify the specific geotechnical and/or geoenvironmental objectives.
4. Create the team.
5. Assess the risk.
6. Determine the most suitable procurement approach.
7. Allocate the risk.
8. Choose the most appropriate contract.
9. Formalise the requirements.
10. Determine the completion criteria.

Figure 2.3 Ten steps to effective planning of ground investigations (continued on next page)

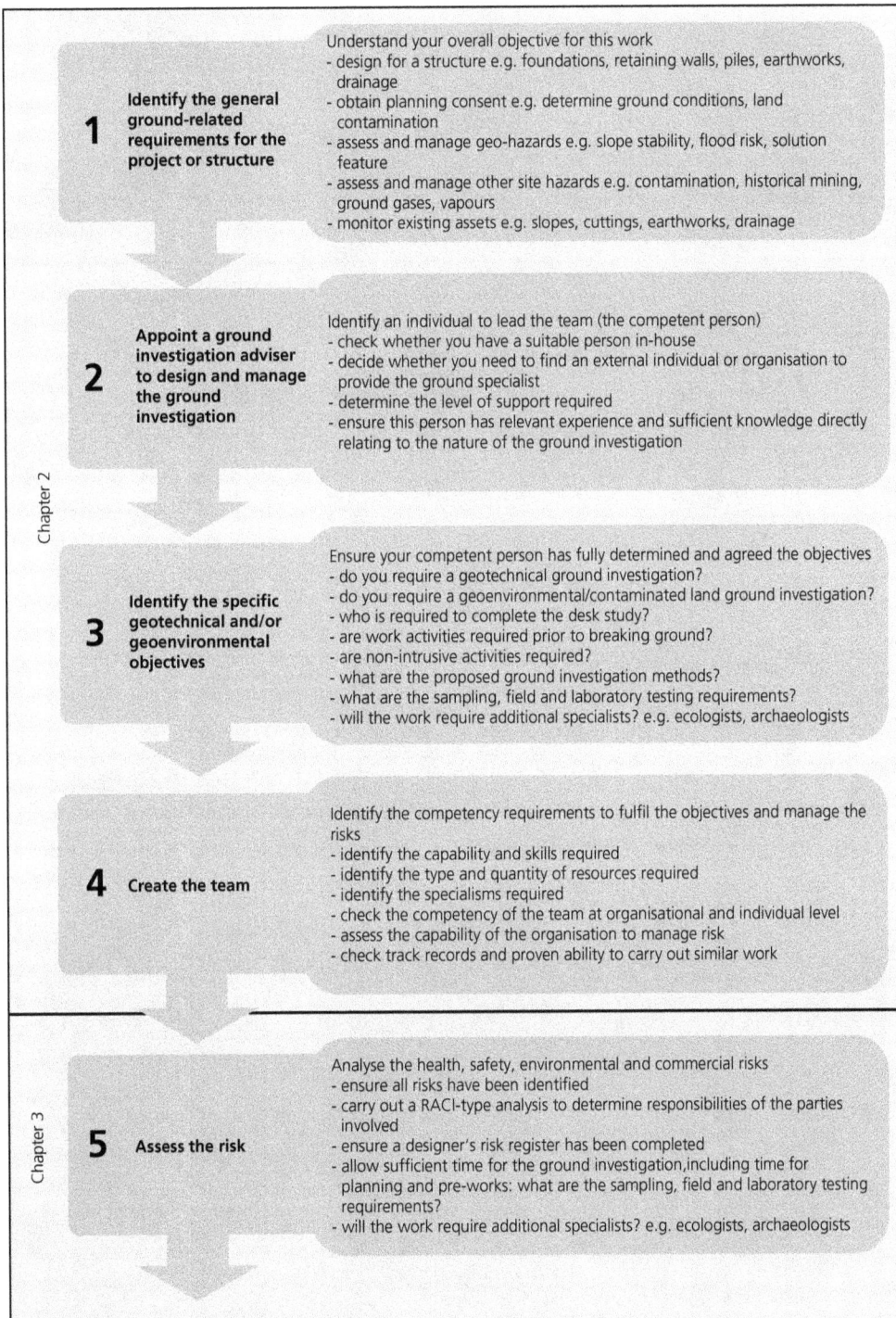

Chapter 2

1 Identify the general ground-related requirements for the project or structure

Understand your overall objective for this work
- design for a structure e.g. foundations, retaining walls, piles, earthworks, drainage
- obtain planning consent e.g. determine ground conditions, land contamination
- assess and manage geo-hazards e.g. slope stability, flood risk, solution feature
- assess and manage other site hazards e.g. contamination, historical mining, ground gases, vapours
- monitor existing assets e.g. slopes, cuttings, earthworks, drainage

2 Appoint a ground investigation adviser to design and manage the ground investigation

Identify an individual to lead the team (the competent person)
- check whether you have a suitable person in-house
- decide whether you need to find an external individual or organisation to provide the ground specialist
- determine the level of support required
- ensure this person has relevant experience and sufficient knowledge directly relating to the nature of the ground investigation

3 Identify the specific geotechnical and/or geoenvironmental objectives

Ensure your competent person has fully determined and agreed the objectives
- do you require a geotechnical ground investigation?
- do you require a geoenvironmental/contaminated land ground investigation?
- who is required to complete the desk study?
- are work activities required prior to breaking ground?
- are non-intrusive activities required?
- what are the proposed ground investigation methods?
- what are the sampling, field and laboratory testing requirements?
- will the work require additional specialists? e.g. ecologists, archaeologists

4 Create the team

Identify the competency requirements to fulfil the objectives and manage the risks
- identify the capability and skills required
- identify the type and quantity of resources required
- identify the specialisms required
- check the competency of the team at organisational and individual level
- assess the capability of the organisation to manage risk
- check track records and proven ability to carry out similar work

Chapter 3

5 Assess the risk

Analyse the health, safety, environmental and commercial risks
- ensure all risks have been identified
- carry out a RACI-type analysis to determine responsibilities of the parties involved
- ensure a designer's risk register has been completed
- allow sufficient time for the ground investigation, including time for planning and pre-works: what are the sampling, field and laboratory testing requirements?
- will the work require additional specialists? e.g. ecologists, archaeologists

Figure 2.3 Continued

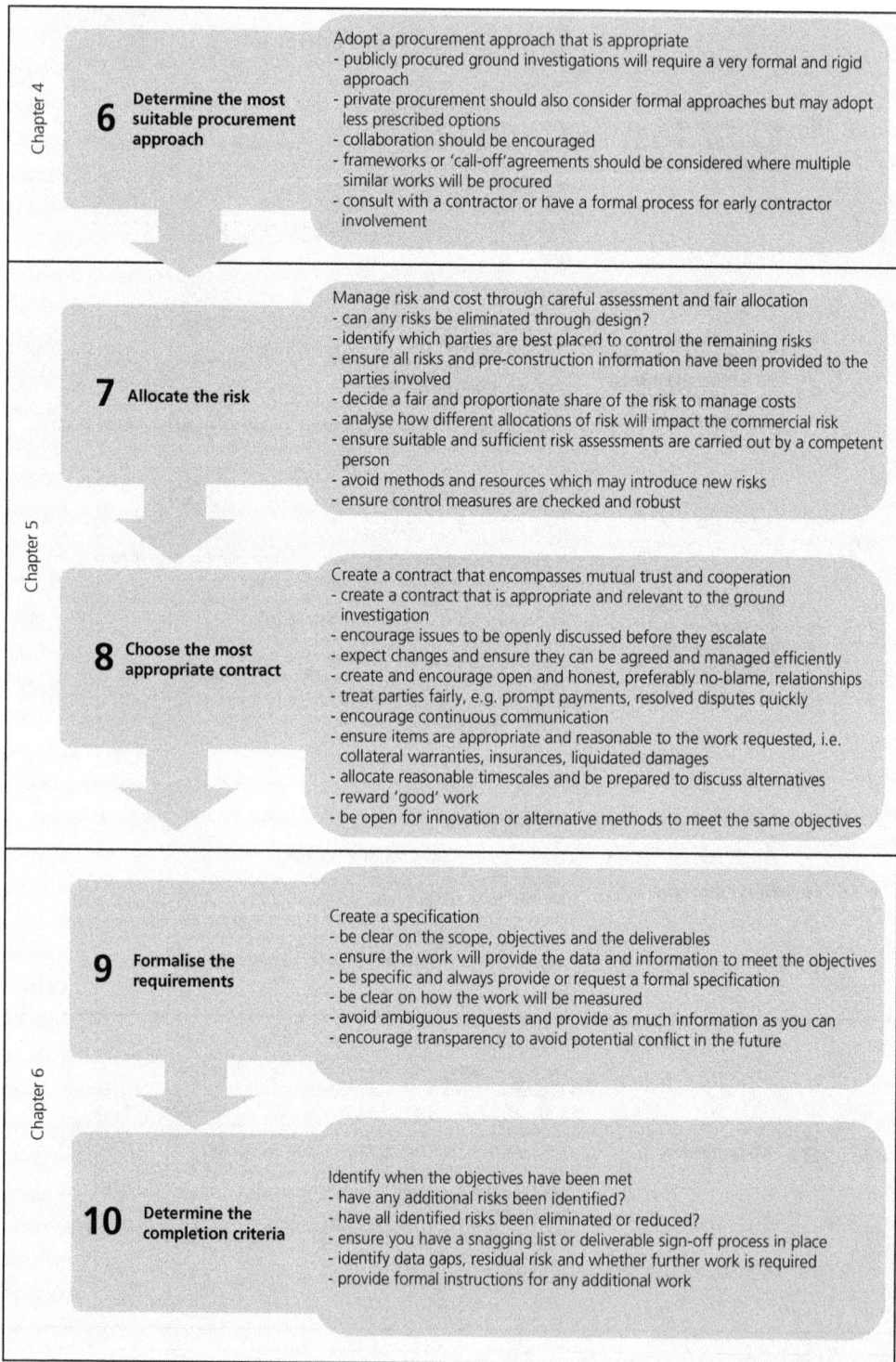

Chapter 4

6 Determine the most suitable procurement approach

Adopt a procurement approach that is appropriate
- publicly procured ground investigations will require a very formal and rigid approach
- private procurement should also consider formal approaches but may adopt less prescribed options
- collaboration should be encouraged
- frameworks or 'call-off' agreements should be considered where multiple similar works will be procured
- consult with a contractor or have a formal process for early contractor involvement

Chapter 5

7 Allocate the risk

Manage risk and cost through careful assessment and fair allocation
- can any risks be eliminated through design?
- identify which parties are best placed to control the remaining risks
- ensure all risks and pre-construction information have been provided to the parties involved
- decide a fair and proportionate share of the risk to manage costs
- analyse how different allocations of risk will impact the commercial risk
- ensure suitable and sufficient risk assessments are carried out by a competent person
- avoid methods and resources which may introduce new risks
- ensure control measures are checked and robust

8 Choose the most appropriate contract

Create a contract that encompasses mutual trust and cooperation
- create a contract that is appropriate and relevant to the ground investigation
- encourage issues to be openly discussed before they escalate
- expect changes and ensure they can be agreed and managed efficiently
- create and encourage open and honest, preferably no-blame, relationships
- treat parties fairly, e.g. prompt payments, resolved disputes quickly
- encourage continuous communication
- ensure items are appropriate and reasonable to the work requested, i.e. collateral warranties, insurances, liquidated damages
- allocate reasonable timescales and be prepared to discuss alternatives
- reward 'good' work
- be open for innovation or alternative methods to meet the same objectives

Chapter 6

9 Formalise the requirements

Create a specification
- be clear on the scope, objectives and the deliverables
- ensure the work will provide the data and information to meet the objectives
- be specific and always provide or request a formal specification
- be clear on how the work will be measured
- avoid ambiguous requests and provide as much information as you can
- encourage transparency to avoid potential conflict in the future

10 Determine the completion criteria

Identify when the objectives have been met
- have any additional risks been identified?
- have all identified risks been eliminated or reduced?
- ensure you have a snagging list or deliverable sign-off process in place
- identify data gaps, residual risk and whether further work is required
- provide formal instructions for any additional work

Creating an effective team

Project size	Large
Informed client	Yes
Geotechnical specialist appointed	Yes
Conditions of contract	ICC

Background

Large-scale infrastructure project across a major UK city requiring an extensive programme of ground investigations. The project involved access to numerous plots of land across several kilometres and often involving permits and formal applications. The project included significant construction and geotechnical challenges including tunnels, earthworks, bridges, portals and so on.

Challenges

- dense urban environment
- complex geotechnical and ground-related structures
- difficult access – multiple landowners, highways, urban and so on
- multiple ground investigation contractors required
- multiple specialisms required

Positive actions

- The Client appointed three framework ground investigation contractors.
- The Client had their own ground investigation advisers to oversee the entire process.
- The Client appointed an independent experienced ground engineering organisation to plan and supervise the ground investigations.
- The Client appointed geological formation experts to provide training and supervision.
- The Client retained a dedicated access team.
- The Client appointed other specialists including unexploded ordnance (UXO), archaeology and geoenvironmental.
- The Client maintained a list of approved subcontractors which could be used by any framework contractor.

Lessons learnt

- The Client recognised the importance of obtaining high-quality investigations from the outset and, therefore, the value of forming an effective, broad-ranging team.
- Appointment of the team early on reduced the Client's risk (less than 2% of claims were attributed to unexpected ground conditions).
- The Client ensured early identification and appointment of subcontractors and other specialists who could be used by all the framework contractors.
- Use of a pool of approved interdependent contractors, subcontractors and specialists ensured flexibility, quality, continuity and consistency.

REFERENCES

ADS (Archaeology Data Service) (2024) https://archaeologydataservice.ac.uk/ (accessed 09/10/2024).

Association of Geotechnical and Geoenvironmental Specialists (2022) *UK Specification for Ground Investigation*, 3rd edn. Emerald/ICE Publishing, Leeds, UK.

Barnes M (1969) Course titled 'Time and Money in Contract Control'.

BSI (2024) BS EN 1997-1:2004+A1:2013: Eurocode 7. Geotechnical design – General rules. BSI, London, UK.

Burland J, Chapman T, Skinner H and Brown M (eds) (2024) *ICE Manual of Geotechnical Engineering*. Emerald/ICE Publishing, Leeds, UK.

Historic England (2020) https://historicengland.org.uk

HEALTH AND SAFETY LEGISLATION:

HSE (Health and Safety Executive) (1974) Health and Safety at Work etc Act.

HSE (1998) Provision and Use of Work Equipment Regulations.

HSE (2006) Personal Protective Equipment at Work (Amendment) Regulations.

HSE (2012) Control of Asbestos Regulations.

HSE (2015a) Construction (Design and Management) Regulations.

HSE (2015b) Managing Health and Safety in Construction: Construction (Design and Management) Regulations 2015: Guidance on Regulations: L153.

FURTHER READING

Geospatial Commission. (2022) Construction Playbook: Why the relaunch will improve re-use of ground investigation data - Geospatial Commission. https://geospatialcommission.blog.gov.uk/2022/09/05/construction-playbook-why-the-relaunch-will-improve-re-use-of-ground-investigation-data/.

HMG (His Majesty's Government) (2022) *The Construction Playbook*. Government Guidance on sourcing and contracting public works projects and programmes. Cabinet Office, OGL, London, UK.

ICE (Institution of Civil Engineers) (2023) *Register of Ground Engineering Professionals (RoGEP)*. ICE, London, UK.

SiLC (Specialist in Land Condition). (2024) Register of professionals maintained by Forum Court Associates. https://www.silc.org.uk/ (accessed 09/10/2024).

emerald
PUBLISHING

ice
Publishing

Association of Geotechnical and Geoenvironmental Specialists
ISBN 978-1-83662-029-7
https://doi.org/10.1108/978-1-83662-026-620251003

Chapter 3
Identifying and managing risk (commercial, health, safety and environmental)

3.1.   Introduction

Clients should recognise that assessment of risk is a fundamental part of the procurement process. It should be undertaken as a collaborative process recognising the skills, knowledge and experience each party brings to the project. Clients should also recognise that risk can be identified at any stage of the project.

An effective ground investigation developed and carried out by an integrated team (including Client, Designer and Contractor) will reduce the risks associated with the ground and groundwater; appropriately qualified and experienced ground practitioners are a key part of such a team. Ground investigation is a process that has its own associated risk elements. By extension, effective procurement of ground investigation will reduce the risk associated with the construction process.

All projects involve an element of risk. This can be commercial risk, where costs might escalate, completion dates become delayed and so on. It might be risk associated with health and safety, where the possibility of injury, illness or ill-health to workers developing the project, users of the project or other people may exist. There can be risk to the environment, to sensitive receptors (humans, animals, plant life, waterways etc.). Risks can be combined: pollution to the air (environmental risk) may then generate fines imposed by the authorities (commercial). Risks can be less obvious: environmental receptors include heritage and archaeology; commercial risks can include reputational damage or costs necessary to protect against hostile acts from activist groups that may oppose the project. Failure to obtain competent professional geotechnical advice also poses a risk to any project by potentially resulting in inadequate ground investigation which may lead to significant programme and cost increases.

The biggest risk for any project which interacts with the ground is the ground itself (Figure 3.1).

3.2.   The risk management process

Consideration of health, safety and environmental factors will reduce the commercial risk to projects and those involved with delivering them, such as impacts to programme, morale, reputation, cost and so on.

When carrying out the ground investigation works, all parties, including the Client, should understand and implement the principles of prevention, as detailed in the Management of Health and Safety at Work Regulations, 1999, which will help ensure that the health, safety and welfare of workers, and anyone who may be affected by the work activities, are protected. The principles of

Figure 3.1 Examples of potential ground related risks (after Baynes, 2010)

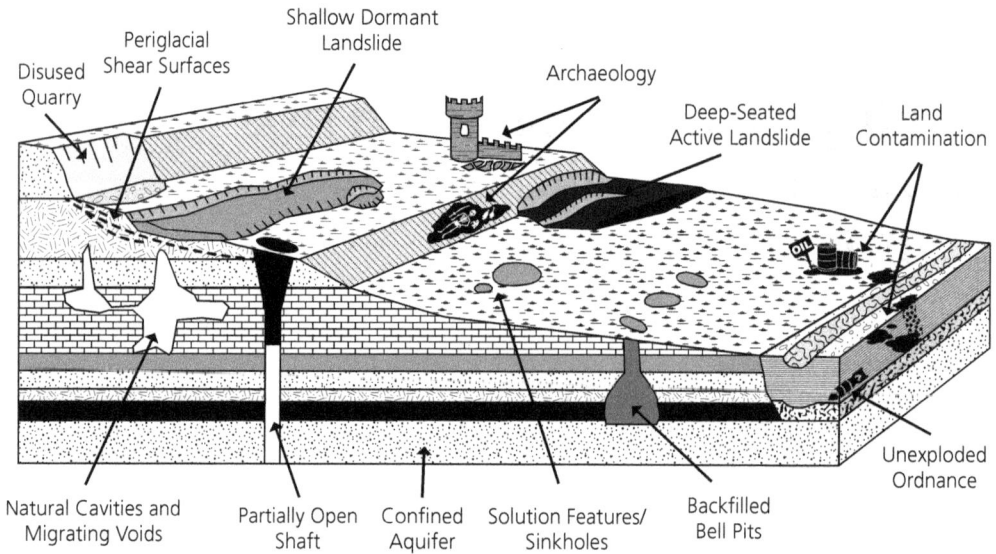

prevention form a best practice approach to risk management, which is summarised in the Construction (Design and Management) Regulations 2015 as

- avoiding risks where possible
- evaluating those risks that cannot be avoided
- put in place proportionate measures that control them at source.

Appendix A provides a description of the nine principles and examples of how Clients may relate these to ground investigations. These principles form a critical foundation to a comprehensive and proactive management system and should be used during the preconstruction and construction phases to develop the best controls for reducing risk, where avoiding risk is not possible.

The principles of prevention are often implemented through a hierarchy of controls which are summarised in Figure 3.2; examples related to ground investigation are provided in Table 3.1.

It is important that this hierarchical approach is understood and applied from the very start of the ground investigation process, but also that it remains the guiding principle throughout the investigations. If conditions change in some way, leading to a new hazard to be addressed, it is not enough to simply 'issue appropriate PPE'; rather a full reassessment of the work is needed, with the aim to eliminate the new hazard if possible.

Risk assessment starts with the Client. For projects where it is foreseeable that more than one contractor will be appointed, under the CDM Regulations the Client must appoint a Principal Designer who will assist in obtaining relevant information about the project and its location, such as past use, existing utilities, UXO risk and so on. The Principal Designer ensures that Designers carry out design risk reduction processes for the design, and documents the actions taken and residual risk on a Designer's risk register (or alternative risk reduction tool). The Designer's risk register details suitable and

Figure 3.2 Hierarchy of controls

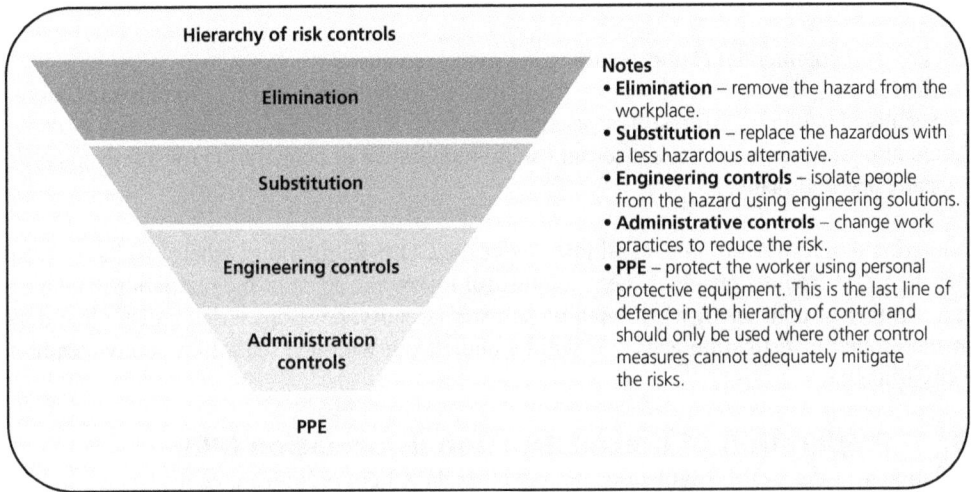

Hierarchy of risk controls

Elimination

Substitution

Engineering controls

Administration controls

PPE

Notes
- **Elimination** – remove the hazard from the workplace.
- **Substitution** – replace the hazardous with a less hazardous alternative.
- **Engineering controls** – isolate people from the hazard using engineering solutions.
- **Administrative controls** – change work practices to reduce the risk.
- **PPE** – protect the worker using personal protective equipment. This is the last line of defence in the hierarchy of control and should only be used where other control measures cannot adequately mitigate the risks.

sufficient mitigation and highlights residual risk generated by the project's undertakings, taking into consideration the life cycle of the project – that is, the construction, maintenance undertaken on the project when complete, those that interact with the structure as a workplace and risk which could impact those demolishing the structure in the future. This information, together with the relevant information obtained by the Principal Designer, forms part of the preconstruction information (PCI).

Table 3.1 Examples of hierarchy controls for ground investigation

Hierarchy control	Example
Elimination	The risk from staff working in a hand-dug trial pit can be eliminated by using a machine to excavate the trial pit and logging from ground surface, removing the need for personnel to work below ground.
Substitution	Substituting handheld sampling and testing equipment by machine-based equipment reduces hand/arm vibration and manual handling – for example, replacing handheld dynamic sampling equipment by a tracked dynamic sampling machine.
Engineering controls	Guarding on drilling machines is a requirement of the PUWER Regulations 1998, preventing personnel coming into contact with dangerous parts of the equipment.
Administrative controls	Requiring staff to use pedestrian walking routes in order to be separated from moving plant is an example of an administrative control. Other administrative controls include policies and procedures for activities. The administrative controls are only effective if staff follow those controls.
Personal protective equipment	Provision of head protection where there is a risk of falling objects; provision of gloves for hand protection. N.B. PPE only protects the person wearing it, while controlling the risk at source protects everybody.

As a Client for a project, there are significant advantages in managing health, safety and environmental risk over and above their legislative obligations. Examples which can be reduced or even eliminated include

- delays and commercial claims resulting in unforeseen costs
- enforcement involvement such as by the police, Environment Agency or Health and Safety Executive leading to unforeseen costs and delays
- reputational damage for the above can lead to withdrawal of cooperation for involved parties locally and nationally.

While ground investigation in general falls under the CDM Regulations, where works do not fall under the CDM Regulations, the Client should still follow the spirit of the regulations and ensure that a member of their team, which may be an external party, develops a risk register and details how the ground investigation activities will be undertaken with due regard for safety, health and the environment.

3.3.   Preparation of preconstruction information (PCI)

Regulation 4 of the CDM Regulations requires that the Client provides preconstruction information (PCI) to every Designer and Contractor appointed or being considered for appointment, to the project. The regulations define PCI as

> information about the project that is already in the Client's possession, or which is reasonably obtainable by or on behalf of the Client.

The regulations state that the PCI must include proportionate information about

- the project
- the planning and management of the project
- the health and safety hazards of the site
- any relevant information in an existing health and safety file.

For a ground investigation, the PCI should also include

- information on the geology of the site, including published information from geological maps and the results from other ground investigations on or near to the site
- information on buried utilities or subsurface structures and obstructions
- information on unexploded ordnance (UXO) risk
- environmental constraints (including contaminated land, asbestos, archaeology, ecology, protected species and invasive plants).

Where some or all of the required PCI is missing, the Client is required to

> take reasonable steps to obtain the information needed to fill any gaps identified by, for example, commissioning relevant surveys.

It may be possible to arrange for this to be done by the Designer or the Contractor; where this is the chosen method, details should be detailed within the PCI, together with the information obtained through these commissioned surveys. The PCI should then be reissued to the project duty holders.

The Client, Principal Designer and other duty holders must regularly review and, if necessary, update the PCI to reflect any changes or developments in the project. This is particularly important as the project progresses and new information becomes available.

Each time the PCI is issued, the Designer will need to consider how the assembled information affects the scope of the ground investigation, while the Contractor will need to consider their arrangements for mitigating risks identified from within the PCI – for example

- identification of buried services (in accordance with PAS 128 (BSI, 2022)
- mitigation of UXO risk (in accordance with CIRIA C681 (Stone *et al.*, 2009)
- management of environmental constraints (ecology survey)
- archaeological watching brief or other appropriate action.

The PCI provided must be relevant to the project and be provided in a timely manner. It should be supplied early enough in the project to allow Designers and Contractors to consider health and safety issues in their pricing, planning and decision-making processes. Understanding the conditions and constraints outlined in the PCI allows for more accurate cost estimation during the preconstruction phase. This helps in developing realistic project budgets and reducing the likelihood of unexpected expenses. Information should be specific, appropriate and relevant to the work being done.

The Client and Principal Designer must ensure that the PCI is accessible to all relevant parties. This includes making the information available to Designers, Contractors, and any other duty holders who require it.

Collaboration and communication are essential. The Client and Principal Designer should actively engage with other duty holders to share information and ensure that everyone has a clear understanding of all health, safety and environmental considerations.

3.4. Management and control of residual risk

The CDM Regulations require the Principal Contractor to produce a construction phase plan (CPP) to demonstrate to the Client that residual risks have been acknowledged and mitigated or controlled. The CPP also confirms that the requirements of CDM Regulations Part 4 – General Requirements for all construction sites have been met. This document, although prepared by the Principal Contractor, should be reviewed by the Client (or Client representative) to ensure alignment with the CDM Regulations. The Principal Contractor must then ensure that either the entire CPP or relevant sections of it is issued to all Contractors who are scheduled to work on the project.

The Client may appoint the ground investigation contractor as Principal Contractor. If the Client chooses not to do so, the ground investigation contractor remains a 'Contractor'. If the Client fails to appoint a Principal Contractor where more than one contractor is appointed, the Client retains the statutory duties of the Principal Contractor themselves.

As defined by the CDM Regulations, whoever is appointed as Principal Contractor must identify hazards that may be present during the work activities and ensure that suitable and sufficient mitigation measures are put in place to prevent harm and that safe systems of work are developed. The construction industry has adopted the term 'risk assessment and method statements' (RAMS) for this process. The *UK Specification* provides guidance on aspects which should be addressed in the site-specific RAMS; further guidance on the risk assessment process can be found in the *AGS Guidance on Risk Assessments and Method Statements for Geotechnical and Geoenvironmental Specialists* (AGS, 2020).

Given the complexity of challenges that ground investigations face in terms of operational risks, a collaborative approach by Clients, Principal Designer, Designers and Contractors when planning the work is essential if risks are to be effectively managed.

It is beneficial to any ground investigation to establish a collaborative environment which encourages the use of methods and approaches which either eliminate or reduce risk to avoid incidents or delays. However, for this to take place, the Client must make provision for the cost of this approach and recognise it as a legal requirement.

The saving made through the absence of robust risk mitigation methods over many years are insignificant when compared to the cost of a single significant accident. When combined with project delays, management time, staff absences, reputational damage and so on caused by minor accidents, the cost of not implementing robust risk mitigation methods can be considerable.

3.5. Consequential hazards

Clients and their Designers should, at the earliest opportunity, involve ground investigation specialists and other stakeholders, such as the end users of the land or structure, to reduce the potential for consequential hazards.

Ground investigation methods can lead to consequential hazards, such as

- creating or destabilising geo-hazards, such as unstable ground, sinkholes or landslips
- causing collapse of mine workings/shafts
- bringing asbestos or other contaminants to the surface
- polluting aquifers or watercourses with arisings and/or contaminated material
- contaminating wastes from site
- causing gas emissions from mine workings or natural materials
- causing vapour emissions from contaminated ground
- causing instability or damage to existing structures
- striking underground utilities and releasing energy, gas, water or sewage
- creating soft and/or uneven ground – for example, access routes and areas churned up by the movement of plant
- causing depressions or holes – that is, collapse or settlement of trial pits and boreholes.

3.6. Adequate time and resources

Proper management of risk requires the Client to allow sufficient time and resources to plan for and mitigate hazards within both the preconstruction and construction phases. A significant proportion of this will be required in the initial planning process, where vulnerabilities in the supply chain should be carefully assessed as part of the tender process and continuously reviewed during the delivery of the project.

The temptation to combine key roles in the delivery must be balanced by the potential disruption to the project should the person or organisation become unavailable or prove to have insufficient resources.

Where access is not available to parts of the site, the Client should consider the advantages of a delay at the start of the project or extending the time for the ground investigation works. While delays are often seen as detrimental to the project, ground investigation delays may not have a significant impact on the overall construction phase. A proactive approach may even introduce additional benefits, such as reducing ground damage, environmental emissions and potential vehicle/pedestrian incidents through fewer vehicle movements; reducing frustration and improving safety by freeing up resources, such as specialist plant or trackway; or reduced environmental impact by avoiding works in sensitive areas at inappropriate times.

3.7.   Specific ground-related risks
3.7.1   Introduction
There are a wide range of specific ground hazards and associated risks that need to be assessed when planning and procuring a ground investigation which, while not common to every investigation, are frequently encountered.

Contractual considerations for work over or near water are discussed in Chapter 5, Section 5.8.3.

3.7.2   Groundwater aquifers and source protection
The presence of groundwater is often a key factor in any engineering project and this applies to the ground investigation stage as well as construction.

Accurately determining the groundwater profile within the ground and how this varies seasonally or under other influences is often one of the main aims of a ground investigation, providing critical information for design. However, it is important to have a sensible expectation of groundwater conditions prior to commencing the ground investigation.

Where intrusive ground investigation is to penetrate an aquifer, it is a common requirement that the ground investigation includes aquifer protection measures to prevent contamination of the aquifer. Where water is being abstracted from the aquifer, there may be restrictions on boring close to the licenced abstraction point.

Conversely, groundwater may already be contaminated; in such cases, consideration should be given to how the health and safety of the investigation workforce are protected, ensuring that new migration pathways that would enable contamination to spread are not created and providing suitable methods of disposal for groundwater – for example, following pumping tests or purging of groundwater instrumentation as part of sampling.

A further consideration is the potential for artesian groundwater pressures to be encountered. These can pose a direct safety hazard to the workforce undertaking the investigation. Beyond that, uncontrolled artesian water can rapidly flood the area of the investigation, potentially leading to significant and costly damage.

Ground investigations also have the potential to affect the groundwater conditions on site. This may be during the investigations themselves – for example, through pump tests – or post investigation, when poorly backfilled boreholes create potential seepage paths for groundwater. Where the ground investigation may affect the existing groundwater conditions, consideration should be given to any possible implications – for example, drawing down of groundwater through a pumping test may lead to ground movement and possibly to structural damage of nearby properties.

The issue of poorly backfilled boreholes or groundwater instrumentation left in place is also a standard concern for tunnelling projects, where encountering such a feature can lead to dangerous water ingress at the tunnel face.

In relation to this, it may be noted that, while it is generally desirable that any groundwater instrumentation installed as part of a ground investigation should be left in situ to provide monitoring capability throughout the construction phase, there are situations where it is necessary to remove instrumentation. The requirement to decommission and remove any instrumentation installed as part of a ground investigation should be identified as early as possible. It is likely that decommissioning will require remobilisation of the boring plant used to undertake the investigation works and appropriate allowance of time and money should be made for such activities.

3.7.3 Archaeology/geoarchaeology

Archaeology is the study of human history and prehistory through the investigation of sites and is a material consideration in planning. The National Planning Policy Framework (NPPF) (MHCLG, 2021) sets out the government's planning policies for England and provides guidance for planning authorities and developers on the conservation and investigation of heritage assets. It considers historic buildings, landscapes and archaeological remains of heritage interest as assets that are afforded statutory protection (specifically dealt with in Section 16 of the NPPF). The primary objective of the NPPF is to foster the delivery of sustainable development not to prevent it, but unique historical information can be lost if sites are destroyed without record. Should archaeological remains be encountered, then these can be preserved 'by record' (a process of mitigation that records and removes remains) or on site by way of design (Historic England, 2016).

Provisions and guidance for protecting assets differ across the UK (NPPF, Historic England, Historic Scotland, Cadw and Northern Ireland Environment Agency) but, generally, consent must be obtained for intrusive ground works in areas of archaeological sensitivity. The sensitivity or not of a site will be described in a desk study during planning (CIfA, 2020b; Historic England, 2015a). A good desk study helps the Client understand ground risk related to archaeology or geoarchaeology by setting out the baseline understanding of any buried heritage, the likelihood of finding artefacts and the heritage significance of any known or potential remains. Heritage risk can be expensive and unexpected discoveries can impact the programme and cost through changes to design and delivery.

Using ground investigation early in the project life cycle can assess and reduce risk by understanding the buried resource and avoiding unexpected discoveries. This offers the first step to managing and protecting heritage by building archaeological requirements into the wider project, forming a targeted strategy or, alternatively, allowing investigation to be descoped.

Ground investigation is particularly relevant to geoarchaeology, using earth sciences to understand the archaeological record (Historic England, 2015b). Ground investigation can be used in predictive geoarchaeological modelling (Historic England, 2020) to reconstruct the past topography and hydrology and indicate the potential nature, depth and location of past human activity as well as deposits likely to preserve environmental remains ('ecofacts'). Alternatively, information from ground investigation can lead directly to the design of intrusive archaeological investigation, such as a purposive borehole campaign or trenching.

Many Designers have heritage specialists within their environmental teams; Clients can check the Chartered Institute for Archaeologists (CIfA) for registered organisations who are practised at providing advice on archaeology within the planning process. Advice should be clear, compliant, impartial, informed and robust. It should be proportionate to a researched and reasoned assessment of the known or potential significance of the heritage on site (CIfA, 2020a). Archaeological input at the ground investigation stage is by no means always necessary. The heritage consultant will work on behalf of the Client to design and procure any required archaeological works related to ground investigation. This involves making appropriate recommendations in the desk study, engaging the local planning authority's (LPA) archaeological advisor to acquire agreement and writing the scope and method (a written scheme of investigation (WSI)).

3.7.4   Natural cavities

Significant areas of the UK are underlain by rocks that have the potential to develop natural cavities. These rocks are principally limestones or rocks with a high calcareous content – that is, chalk. Solution of these rocks by groundwater leads to the formation of cavities that can be small and localised, ranging to voids that are extensive in both height and lateral extent. Where such rocks are known or suspected from the desk study, more specific information should be sought from the British Geological Survey (BGS) and from local councils.

Some natural cavities may also have been caused by glacial activities – that is, drift filled hollows and infilled chalk solution features. There is also the potential for natural cavities to have been backfilled by anthropogenic processes.

It may be necessary to engage the services of a specialist who has specific knowledge of geohazards within the proposed development area.

3.7.5   Mining and mineral extraction

Large parts of the UK are underlain by mineral workings. While many of these relate to coal extraction, substantial underground voids were created by the extraction of limestone, sandstone, salts and other mineral types. As part of the risk control, Clients should seek to identify the extent of these workings by ensuring that they appoint an appropriate specialist – for example, local geological expert, mining specialist and so on – and produce a comprehensive desk-study mining risk assessment and mine gas risk assessment.

While there may be good records of some mining activities, particularly those that are more recent, the history of mining stretches back at least 6000 years to the Neolithic period. As such, abandoned workings, including shafts and adits, may be unrecorded, so it is important to have an awareness of any past mining activity in the area, even if there is no record of specific activities on the site under investigation.

Guidance on undertaking construction works in the vicinity of old mine workings may be obtained from *CIRIA C758 Abandoned Mine Workings Manual* (Parry and Chiverell, 2019), which features extensive guidance on desk studies and ground investigations in areas of mine workings.

In areas where coal mining has been undertaken, works that are likely to interface with coal workings will need permission from the Coal Authority; specifically, a permit is required to drill boreholes through a coal seam. The Coal Authority also maintains publicly accessible records on past coal extraction activities although, as noted above, these cannot be considered to be a complete record of all such activities.

Consideration should be given to less widely known organisations that have responsibility for mining or mineral extraction activity – for example:

- Mining at a small scale occurs in the Royal Forest of Dean; records are held by the Deputy Gaveller of the Forest of Dean.
- The Cheshire Brine Subsidence Compensation Board deals with claims for compensation related to subsidence from the pumping of brine under Cheshire; however, it has no responsibility for maintaining records of current or abandoned mines.

Where ground investigations are to be undertaken in an historical mining area, it is particularly important that this is recognised during the desk study stage. An understanding of the geology and history of the area may identify the likelihood of historical mining activity. Historical mapping can provide useful clues to the presence of mining or brine extraction – for example, abandoned shafts or ground settlement leading to the formation of ponds.

There is a range of hazards associated with past or current mining activity that can affect ground investigation works and which can be triggered or exacerbated by such activities. The most obvious of these is the potential to trigger collapse, which can pose a direct threat to life of the investigation workforce if occurring during the investigation, or future land users if the investigation destabilises the workings.

Ground investigations have the potential to cause heating of any coal remaining in the ground and have been responsible for starting subsurface fires.

Mines may contain acidic or metal-contaminated groundwater or hazardous gases/atmosphere, which can pose a direct hazard to personnel undertaking the investigation. Further, if the investigation opens new pathways for migration, such water or gas can affect other receptors – for example, gas migration into buildings, polluted groundwater leaking into a drinking water aquifer.

Figure 3.3 Ground investigation and grouting of identified mine workings. Courtesy of Chris Wade, Soil Engineering Geoservices Limited (AGS photography competition entry)

Much mineral extraction has been undertaken by opencast workings/quarrying. Backfilling of such works may have been carried out to engineering standards, but many have been executed poorly. The abandoned pit may have been used for general waste dumping, so there are potential hazards associated with land instability, highly heterogeneous ground and contaminated ground. These can pose a direct hazard to the investigation personnel and lead to delay and additional cost if they were not allowed for during the planning of the investigation.

Mining areas often contain numerous spoil heaps; the stability of the spoil heap is a potential hazard to consider regarding any work done in its proximity. Spoil from mines may have been used as general fill and may have concentrations of metals, hydrocarbons and so on that require the site to be treated as contaminated land.

A further consideration is whether an investigation is planned for a location with current or potential future mineral extraction. There may be a need to obtain permission from whoever holds the licence for mineral extraction. If located in an area where there are active protests against mineral extraction, such as fracking operations in areas of oil/gas containing shales, there is the possibility that a ground investigation rig could be mis-identified and subject to vandalism/attack.

3.7.6 Ground gas and hazardous atmosphere

There are a variety of situations that may arise where, in procuring a ground investigation, the potential for gas or other atmospheric hazards should be considered. Ground investigations are typically undertaken from the ground surface in external environments, so are generally in well-ventilated areas, and the hazard posed by ground gas or vapour during drilling (if it occurs) can normally be easily managed with simple precautions.

Works may also be undertaken in a basement or existing tunnel, for example, where confined space working conditions (as defined by the Confined Space Regulations 1997: HSE, 1997) apply. Even in areas that are not legally defined as confined spaces, conditions of ventilation may need to be assessed. The use of standard diesel-fuelled plant may, for example, be inappropriate if ventilation systems to remove the fumes cannot be established, leading to a requirement to use electrically driven plant. Gas emissions from boreholes inside buildings, if high enough, can create hazardous atmospheres or displace oxygen, creating an asphyxiation risk unless appropriate ventilation is established.

There may be additional atmospheric hazards that should be considered at time of procurement. Fumes or dust from industrial processes on or about the site may generate short-term health issues (headache, nausea); prolonged (day-long) exposure may pose longer-term health risks. Fumes may be potentially explosive (e.g. petrol station forecourts, refineries, fuel storage locations etc.), placing limitations on the types of plant and method of working.

Gas concentrations from a variety of sources can accumulate at hazardous levels around the top of boreholes during drilling if pockets of gas are encountered that are under pressure. Gases can occur within or adjacent to alluvial soils, landfill sites, hydrocarbon source rocks, coal measures, mine workings and where there is deoxygenated air present in the Lambeth Group in London. The area affected around the top of the borehole is normally small and emissions usually reduce quickly within a few hours as the pressure reduces. Carbon dioxide can also accumulate in the bottom of trenches, and this is a known risk in excavations in former mining areas and in colliery spoil.

While the likelihood of gas at a volume and/or pressure that poses a hazard being encountered is normally small, the associated risks need to be assessed for each site.

Beyond the risk posed to the investigation workforce, investigations seeking to establish conditions related to ground gas or vapour must be designed with this in mind. The desk study and ground model for a site should determine whether gas monitoring is necessary (refer to *Hazardous Ground Gas Top Tips* (SoBRA, 2023) and *Hazardous Ground Gas – An Essential Guide for Housebuilders* (NHBC, 2023)).

Reference should be made to BS 8576 (BSI, 2013) for more details of the investigation for ground gas and vapour monitoring. This includes important guidance on monitoring well installation, gas monitoring methods and how monitoring should be carried out.

It is important that the investigation works themselves do not form a pathway for gas/vapour migration that then becomes a hazard to site uses after the investigation is completed. Boreholes and monitoring wells should be backfilled or decommissioned in accordance with Environment Agency or SEPA guidance.

3.7.7 Potentially contaminated sites, atypical and emergent contaminants

While ground investigations routinely deal with the risks associated with contaminated soil and groundwater, potential contaminants still need to be identified and properly assessed for their potential to do harm to both those involved in the investigation and any other third party. Any investigations also need to include measures to ensure that the investigation does not spread contamination or create new migration pathways for contaminants. The British Drilling Association (BDA, 2024) have provided guidance on *Safe Intrusive Activities on Land Potentially Impacted by Contamination*, while British Standard BS 10175: *Investigation of potentially contaminated sites. Code of practice* (BSI, 2011) provides specific guidance on the investigation process.

Figure 3.4 Ground investigation involving potential contamination from buried tanks. Courtesy of Alan Beattie, Terra Firma (Wales) Ltd (AGS photography competition entry)

When undertaking investigations in made ground, consideration should be given to the hazard posed by any soil samples that may be sent for geotechnical laboratory testing to the laboratory staff undertaking the tests. Testing laboratories may require screening tests of the sample to confirm that it is safe to undertake the geotechnical testing – for example, the possible presence of asbestos fibres. Geotechnical testing of made ground should only be undertaken where there is a clear requirement for the resulting data.

3.7.8 Unexploded ordnance (UXO)

The UK's history means that the hazard posed by UXO can be significant. This hazard comes from routine operations by the military – it is estimated that over 20% of the UK's land area has been used for military training (CIRIA, 2009) – but also from acts of war directed at the UK, most commonly aerial bombardment carried out during World War II (although there was also aerial and naval bombardment during WWI and long-range shelling during WWII).

CIRIA C681 (Stone *et al.*, 2009) provides guidance on how to address the risks associated with UXO; further guidance is given in CIRIA C785 (Bowman *et al.*, 2019). It should be remembered that UXO hazard applies not just to the development project for which the ground investigation is being carried out, but also to the ground investigation itself. Clearly, it would generally be more efficient to ensure any UXO risk assessment takes account of all stages of the project and the activities to be undertaken in each stage. However, since the potential for UXO penetration into the ground depends to some degree on ground conditions, the ground investigation may provide key input into a revised UXO assessment for the construction stage.

When assessing the requirements for a ground investigation at the procurement stage, it is important not to assume UXO risk applies only to the larger cities that were heavily bombed: rural locations may have less obvious reasons for being at high UXO risk. Some examples of possible less obvious UXO sources include the following.

- RAF airfields that have since been returned to use for farming may have been subjected to bombing or have demolition mines ('pipe mines') placed that might not have been removed.
- Postwar demilitarisation saw items of explosive ordnance being scrapped to recover valuable materials (i.e. metal content). Such sites can have a legacy of waste materials that may not be free from explosives.
- Bombing decoy sites, such as Starfish sites (and similar), were set up during World War II at various locations around the country to protect cities or other valuable targets by creating false targets for bombing missions. These were located in rural areas and were generally returned to agriculture rapidly at the end of the war; there is often poor awareness of their existence, let alone their precise location (see Dobson, 2013).

3.7.9 Work over or near water

Work over and near water includes works that are on or near

- rivers, lakes and other inshore bodies of water
- ports, harbours and bridge structures
- intertidal zones and nearshore.

Figure 3.5 Overwater ground investigation from a floating pontoon. Courtesy of Arie Zamler, Stantec (AGS photography competition entry)

Offshore works are outside the scope of this guidance, since it is assumed that such works are typically more complex and that the parties involved in such works are likely to have the specialist knowledge needed.

When procuring ground investigations over or near water, it is imperative to thoroughly evaluate and address additional safety, environmental and access considerations and to ensure members of the team have relevant experience in this environment and the associated risks.

In order to manage the risks associated with work over or near water, preconstruction information will need to be more extensive than is typical for a land-based investigation – for example, there may be a need for up-to-date bathymetric surveys, risk assessments for UXO, proposed assembly sites, daily access and crew change locations, as well as historical river level or tide and chart data. The specific location of the exploratory holes and investigation methods should be evaluated to ensure that additional risks are not created, such as the potential to pollute a water body with arisings or uncontrolled flush.

3.7.10   Works on or about transport infrastructure

Ground investigation works on existing railway, highway and other transport infrastructure sites come with their own unique challenges that, if not considered during the planning stages of the ground investigation, can significantly impact final delivery and costs. Typically, works need to be planned significantly in advance, with access booked for a limited time window, which may require coordination with other works or activities that need to take place on the transport infrastructure; adequate time and resource should be allocated to organise and book the appropriate access.

Local knowledge of the planning process will provide efficiencies and benefits as each infrastructure operator and, in some cases each region of an operator, employs a different process;

knowledge of planned works may allow certain activities to be undertaken within a worksite booked by others.

Specialist plant, such as road rail vehicles (RRVs), or temporary traffic management, may be required, which can require additional cost and time to deploy.

Should works overrun, with delays impacting the ability for the 'hand back' of the works site to the planned operation of rail services, highway lane reopening and so on, the project may incur substantial delay costs.

3.8. Allocation of risk
3.8.1 Introduction
Allocation of risk should be clearly determined by the form of contract used to procure the ground investigation. Failure to unambiguously assign all risks between the contracting parties poses a significant risk of later contractual disputes.

While there are various forms of contract that may be used for ground investigations (see Chapter 5), NEC4 is widely used.

The parties to an NEC4 contract are

- Client (previously Employer)
- Contractor

with specific duties allocated to the following identified roles

- Project Manager
- Supervisor

Possible additions by way of Z clauses might include Designer and Investigation Supervisor, the latter who may be interchangeable with the NEC 4 Supervisor role.

3.8.2 The New Engineering Contract (NEC) approach
NEC (2017) offers a range of options for payment and delivery models that allow risk to be allocated to the parties, appropriate to the complexity and scale of the project, such as

- straightforward work, where a supplier is willing and able to manage project risk: Clients can let projects using a fixed price contract and benefit from cost certainty
- complex or larger projects where Client and supplier are willing to pool risk: Clients deploy a target cost contract, which incorporates a pain/gain mechanism for sharing any savings or overspends
- projects where the Client believes it is best placed to manage risks: Clients use a cost-reimbursable contract, in which suppliers receive their costs plus an agreed fee, or a management contract, in which suppliers are paid a fee for managing the sub-contractors delivering the works.

The NEC contract in particular requires and encourages a collaborative mindset with good risk management and an emphasis on 'no surprises', relying on excellent communication between the contract parties. The clear allocation of risk from the outset will assist in the management of the contract and allow 'compensation events' (contract price changes) to be more easily recognised and agreed.

Appendix B provides guidance in the form of a RACI (responsibility/accountability/consulted/informed) table on typical tasks and ownership/risk allocation within a ground investigation project. Depending on the contract used, some of the terminology may change and contract-specific requirements may also result in changes to the eventual allocation of risk. However, the items identified here should provide a good starting point for any contract and can be modified or added to suit the particular contract requirements.

3.9. Geotechnical baseline report (GBR)

A GBR is a contract document that describes a set of defined ground conditions. It is different from a ground investigation report or a geotechnical design report, which are defined in British Standards (BS) and Eurocode 7 geotechnical design standards. GBRs based on all information (desk study, nonintrusive and intrusive investigation etc.) will provide essential data to allow suitable risk assessment and management procedures to be adopted

GBRs are principally used during the construction phase of a project and are seeing increasing use throughout the construction industry (CIRIA C807: Davis *et al*, 2023). These are not the direct output from a ground investigation but, where there is the intention to use a GBR on the project, it is important that this be identified as soon as possible, and before the ground investigation is procured.

If the project will use a GBR for construction contract procurement, several additional considerations are necessary when planning the ground investigation.

- A good GBR requires a thorough understanding of the ground conditions to be encountered and what might be encountered – through a combination of ground investigation, field mapping and desk study.
- Ground investigations are typically scoped by Designers with a focus on determination of parameters for design – thinking ahead to the GBR, the ground investigation should also address the key material properties and variability that will impact construction method, plant selection and productivity rates (a comprehensive and well-scoped ground investigation should do this anyway).
- The most onerous conditions for design can sometimes be at odds with the most onerous conditions for construction; for the GBR there should be focus on defining both sets of conditions in the ground investigation.
- A GBR based on a comprehensive ground investigation will provide an effective risk allocation tool.
- The timing of the ground investigation should consider the level of maturity that is required to support the writing of the GBR which must be agreed between parties at time of contract award. Therefore, while the GBR can be used to complement the ground investigation and fill gaps in knowledge at early tender stages, the ground investigation should be as complete as possible sufficiently in advance of contract award to allow the final GBR to be prepared.

Collaboration and conflict avoidance

Project size	Large
Informed Client	Yes
Geotechnical specialist appointed	Yes
Conditions of contract	NEC3

Background

On a large infrastructure project in the UK, the site was subject to some isolated acts of vandalism carried out on plant and equipment, even with security measures in place. The main site compound for one of the phases of ground investigation became a regular target for disruption and picketing when local interest groups engaged with national protester groups who had moved into land adjacent to the site.

Challenges

- Daily blockades of the site entrance required frequent visits by the police and caused significant delays to the programme.
- Information will stay in circulation on the internet even when disproved as fake news.
- Protesters tried to provoke clashes with the intention of filming them for distribution online; people even attempted to cause injuries to themselves by deliberately being in too close proximity to moving vehicles.
- Staff working on the site became worried when protesters began to take photographs of vehicle registrations and ask for personal details online.
- Companies involved in the project were named with a view to protesting at their business addresses.

Positive actions

- Early interaction with opposition was limited to discussions with local stakeholders during negotiations for land access and subsequent works.
- Principal Contractor was forced to apply for an injunction excluding protest from the site.
- Client had to take possession of the protest camp land earlier than planned to evict the camp.

Lessons learnt

- The potential for social unrest and project protests were not sufficiently considered during planning.
- Ground investigation is often one of the earliest and usually the first substantial construction activity on site and therefore is often a likely target.
- Investigation works may attract the interest of local stakeholders, individuals or dedicated protesters that are opposed to the overall project.
- Security measures and wider implications around opposition to works need to be considered as part of the procurement process.

CASE STUDY

Environmental risk

Project size	Large
Informed Client	Yes
Geotechnical specialist appointed	Yes
Conditions of contract	NEC3

Background

On a large infrastructure project, land access negotiations to create sufficiently large packages of work to mobilise the ground investigation contractors to site were protracted due to numerous stakeholders. This meant that the works started later in the calendar year than envisaged.

Challenges

- Specific legally binding commitments were included in the commitment of the Client to the project.
- Surface conditions changed due to time of year, requiring additional trackway to access locations.
- Protected species, great crested newts, used the trackway to hibernate under.
- Special systems of work to mitigate the risk created by changes in ecological or access conditions were required.
- Delays meant that crops were now at an advanced stage of growth when works were to be undertaken.

Positive actions

- Comprehensive ecological surveys were completed at the start of the project (but became obsolete due to the delays).
- Access routes had been chosen and agreed for the original timing of the ground investigation (but conditions changed due to the delays).

Lessons learnt

- The risks caused by a delay to the programme must be assessed by the Client prior to tender, especially those associated with seasonal weather changes, protected species, ecology and surface ground conditions.
- Protected species (fauna or flora), crops and so on may dictate the specific time of year the work can be undertaken.
- Commercial tender processes need to ensure that consideration is given to calendar-sensitive issues.
- Farmers and landowners may be more cooperative in allowing the work at certain times of year.
- Special systems of work to mitigate the risk which would not normally be included in the price and/or programme may be required.
- Clear explanation of any specific requirement the Client has agreed to must be provided at an early stage so that the ground investigation contractor can properly allow for them in their tender, programme and project resources.

REFERENCES

AGS (Association of Geotechnical and Geoenvironmental Specialists) (2020) *AGS Guidance on Risk Assessments and Method Statements for Geotechnical and Geoenvironmental Specialists.* AGS, Faversham, UK.

BDA (British Drilling Association) (2024) *Safe Intrusive Activities on Land Potentially Impacted by Contamination.* Emerald Publishing, Leeds, UK.

Bowman R, Davies P and Baptie P (2019) *Unexploded Ordnance (UXO) Risk Management Guide for Land-Based Projects. CIRIA C785.* CIRIA, London, UK.

BSI (2011) BS 10175:2011+A2:2017: Investigation of potentially contaminated sites. Code of practice. BSI, London, UK.

BSI (2013) BS 8576: Guidance on investigations for ground gas – Permanent gases and volatile organic compounds (VOCs). BSI, London, UK.

BSI (2015) BS 8485:2015+A1:2019: Code of practice for the design of protective measures for methane and carbon dioxide ground gases for new buildings. BSI, London, UK.

BSI (2022) PAS 128:2022: Underground utility detection, verification and location – Specification. BSI, London, UK.

CIfA (Chartered Institute for Archaeologists) (2020a) *Standards and Guidance for Commissioning Work or Providing Consultancy Advice on Archaeology and the Historic Environment.* CIfA, Reading, UK.

CIfA (2020b) *Standards and Guidance for Historic Environment: Desk-Based Assessment.* CIfA, Reading, UK.

CIRIA (2009) *C681 Unexploded ordnance (UXO): a guide for the construction industry.* K Stone, A Murray, S Cooke, J Foran and L Gooderham (Authors).

CL:AIRE (Contaminated Land: Applications in Real Environments) (2012), *A Pragmatic Approach to Ground Gas Risk Assessment.* CL:AIRE, Reading, UK, Research Bulletin RB17.

Davis J, Essex R, Farooq I and Drake A (eds) (2023) *Geotechnical Baseline Reports: A Guide to Good Practice.* CIRIA (Construction Industry Research and Information Association), London, UK, CIRIA C807.

Dobson C (2013) *Fields of Deception.* Methuen, York, UK.

Historic England (2015a) *Managing Significance in Decision-Taking in the Historic Environment. Historic Environment Good Practice Advice in Planning: 2.* Historic England, Swindon, UK.

Historic England (2015b) *Geoarchaeology. Using Earth Sciences to Understand the Archaeological Record.* Historic England, Swindon, UK.

Historic England (2016) *Preserving Archaeological Remains: Decision-Taking for Sites under Development.* Historic England, Swindon, UK.

Historic England (2020) *Deposit Modelling and Archaeology. Guidance for Mapping Buried Deposits.* Historic England, Swindon, UK.

Historic England (2022) *Planning and Archaeology: Historic England.* Historic England, Swindon, UK, Advice Note 17 (HEAN 17).

HSE (Health and Safety Executive) (1997) Confined Space Regulations.

MHCLG (Ministry of Housing, Communities and Local Government) (2021) National Planning Policy Framework. The Stationery Office, London, UK.

NEC (New Engineering Contract) (2017) *NEC Contracts.* NEC, London, UK. https://www.nec-contract.com/ (accessed 09/10/2024).

NHBC (National House Builders Council) (2023) *Hazardous Ground Gas – An Essential Guide for Housebuilders.* NHBC Foundation. Milton Keynes, UK, NF94.

Parry DN and Chiverell CP (eds) (2019) *Abandoned Mine Workings Manual.* CIRIA, London, UK, *CIRIA C758.*

SoBRA (The Society of Brownfield Risk Assessment) (2023) *Hazardous Ground Gas Top Tips*. SoBRA. https://sobra.org.uk/ (accessed 09/10/2024).

Stone K, Murray A, Cooke S, Foran J and Gooderham L (2009) *Unexploded Ordnance (UXO): A Guide for the Construction Industry*. CIRIA, London, UK, CIRIA C681.

FURTHER READING

ADS (Archaeology Data Service) (2023) https://archaeologydataservice.ac.uk/help-guidance/instructions-for-depositors/files-and-metadata/#What%20to%20Deposit (accessed 22/12/2023).

AGS (2022) UK Specification for Ground Investigation, 3rd edn. Emerald/ICE Publishing, Leeds, UK.

BSI (2004) BS EN 1997-1:2004+A1:2013: Eurocode 7. Geotechnical design – General rules. BSI, London, UK.

BSI (2020) BS 5930:2015+A1:2020: Code of practice for ground investigations. BSI, London, UK.

HSE (1998) Provision and Use of Work Equipment Regulations.

HSE (2006) Personal Protective Equipment at Work (Amendment) Regulations.

HSE Books (2015) Managing Health and Safety in Construction: Construction (Design and Management Regulations 2015. Guidance on Regulations L153.

Association of Geotechnical and Geoenvironmental Specialists
ISBN 978-1-83662-029-7
https://doi.org/10.1108/978-1-83662-026-620251004

Chapter 4
Procurement approaches

4.1. Introduction

Effective procurement forms a crucial component of any organisation's supplier management processes, allowing an organisation to operate efficiently to improve quality, while achieving cost savings and reducing risk. It ensures the availability of resources in the necessary quantities, at prices that align with organisational needs and strategic objectives, through a process that complies with legislative requirements and ethical standards.

Procurement is defined by the Chartered Institute of Procurement and Supply (CIPS, 2023) as

> the process of acquiring goods, services, or works from an external source… It involves all the activities required to obtain goods and services, including identifying needs, tendering, evaluating suppliers, negotiating contracts, and managing supplier relationships.

Figure 4.1 provides a suggestion for a series of steps a Client can follow to aid the procurement process.

The arrangements for procuring the suppliers for a ground investigation need to be determined before the procurement process can be commenced. There are several questions that need to be addressed at this stage, not least which party will undertake the procurement.

Supplier procurement might be by the Client, the Designer or the Principal Contractor. Where multiple suppliers are required to deliver the project, consideration is required as to whether a single procurement is appropriate, whether different services are subcontracted through a single supplier or whether individual services are procured separately. While there may be benefits in procuring specialist services separately by retaining control over the suppliers and reducing costs, this may be offset by an increased requirement to coordinate between the multiple suppliers.

On projects that will require multiple phases of investigations over a prolonged period, there may be a benefit in establishing a framework or term contract arrangement with one or more suppliers. The benefits of this approach may include

- reducing the need for repeated potentially time-consuming and costly procurement processes
- the opportunity to develop long-term relationships with suppliers, improving quality and consistency.

Figure 4.1   Seven steps to procurement

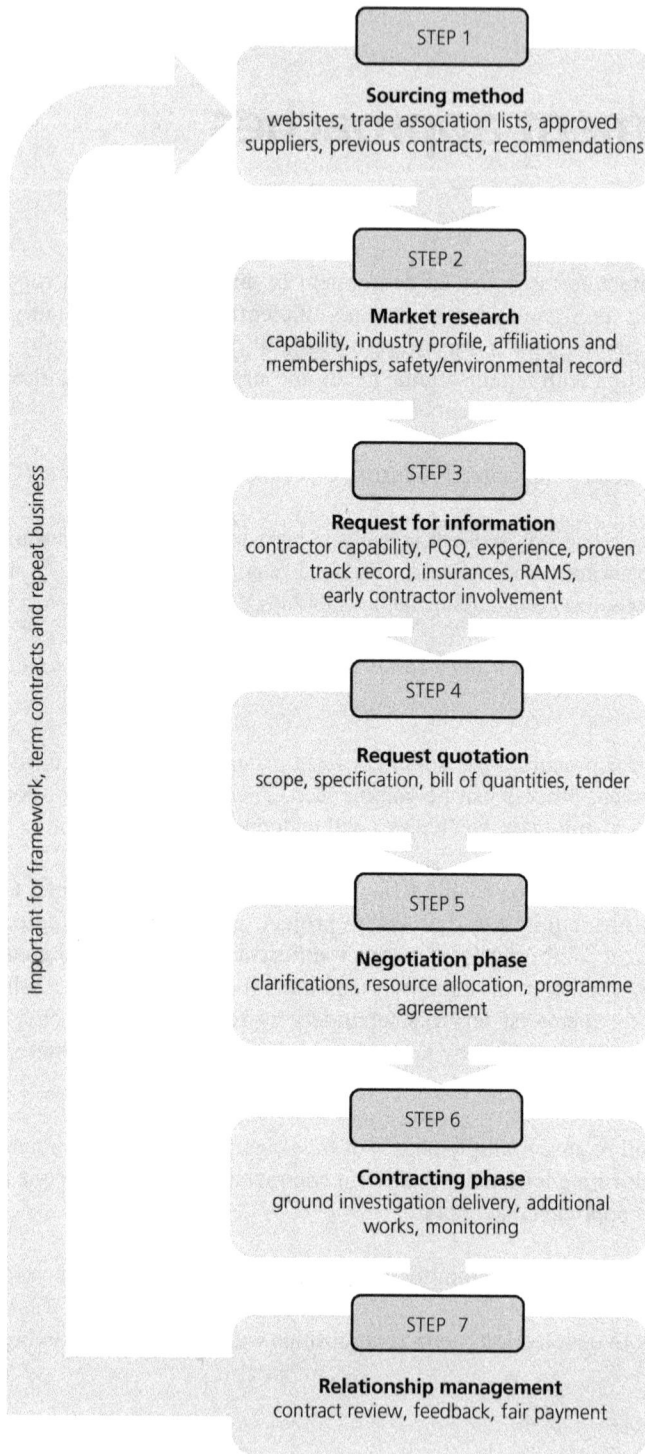

STEP 1

Sourcing method
websites, trade association lists, approved
suppliers, previous contracts, recommendations

STEP 2

Market research
capability, industry profile, affiliations and
memberships, safety/environmental record

STEP 3

Request for information
contractor capability, PQQ, experience, proven
track record, insurances, RAMS,
early contractor involvement

STEP 4

Request quotation
scope, specification, bill of quantities, tender

STEP 5

Negotiation phase
clarifications, resource allocation, programme
agreement

STEP 6

Contracting phase
ground investigation delivery, additional
works, monitoring

STEP 7

Relationship management
contract review, feedback, fair payment

Important for framework, term contracts and repeat business

However, there may be disadvantages associated with this approach, including

- a potentially more complex initial procurement process
- the need to procure suppliers before the scope is fully defined, introducing pricing uncertainties
- the need to establish mechanisms to manage a breakdown in the relationship between supplier, Client and other stakeholders.

Supplier procurement should also take into account the potential benefits of early contractor involvement (ECI). Chapter 4, Section 4.4 provides further guidance on ECI.

Having determined who will be responsible for procurement, it is necessary to establish what form of contract will be utilised for each of the services involved in the investigation (see Chapter 5).

When following a tendering process, consideration should be given to the appropriate balance of price and quality. A weighting of 60% or 70% allocated to quality and safety submissions allows the tenderer to be incentivised to prioritise quality in preference to a 'low quality, low price' submission, with the remainder allocated to the price of the works. A Client may alternatively choose to exclude the lowest price tender as this bid is likely to have compromised on quality and/or time (see Chapter 2, Section 2.2). The assessment of quality of the submission should also include consideration of the proposed supervision of the investigation works, taking into account what party or parties will be responsible for providing supervision of the investigation.

It may also be necessary to establish whether there are any legal or other constraints on procurement, particularly in relation to public sector projects. Public sector Clients are likely to have constraints on the procurement of suppliers, requiring an appropriate competitive tender process to be followed, whereas private sector Clients may have fewer constraints, although a competitive tender process should still be beneficial to achieve value for money while still achieving the objectives.

Due to UK legislation, the needs and requirements of the procurement approach are determined by the type of Client organisation and the purpose of the goods (or supplies), services or works that are to be acquired. Specifically, it depends on whether those acquisitions are required on behalf of a public, or non-profit, sector organisation or company, or whether they are for the private sector.

In the public sector, procurement activity is regulated. The regulations are designed to ensure that public funds are spent in a fair and equitable manner and are transparent to all. In the private sector, procurement is largely unregulated and organisations are free to procure their own procurement strategies.

4.2. Collaboration

The purpose of a ground investigation is typically to de-risk later stages of a project where costs can be significant. An effective ground investigation has the potential to enable efficient design, reduce project timelines and save construction costs. To achieve this aim, a ground investigation should focus on outcomes – that is, the delivery of high-quality data and not price. The focus on a Designer's insurance levels or liquidated damages that should be applied to a Contractor may not be in the best interests of a project if revealed ground conditions merit a different approach or further investigation.

A successful ground investigation (whether procured within the public or private sector) should, therefore, draw on the collective experience of all parties in terms of collating site knowledge and deploying the most appropriate methods of investigation at both design stage and delivery of the subsequent ground investigation.

Project stakeholders may not be the best placed to manage ground risk. Designers and Contractors are likely to add significant value by deploying experienced staff, local knowledge and industry best practice. As discussed above, on larger projects this can be achieved through formal contractual arrangements such as an ECI stage. However, collaboration should not be dependent on project value or contractual arrangements.

Collaboration can foster tangible benefits for a project. Consideration and discussion of the practicalities of delivery (such as considering access routes, extent of enabling works, seasonal constraints) at an early stage may allow cost savings to be made and development of a better value proposition. Discussion of the delivery of other advance survey works (e.g. geophysics, archaeological, ecological or other surveys) in a single campaign may also reduce impact on landowners and other stakeholders while providing a more holistic understanding of the site.

However, while cooperation should be fostered during the initial design and mobilisation stage, it should continue through delivery and be maintained in the production of final deliverables. This is achieved by ensuring the timely capture and effective dissemination of useful information for the mutual benefit of all parties.

Parties should be cognisant of the barriers to this collaborative effort. By definition, often little pre-existing information is available for a site and there is a risk of unforeseen conditions resulting in a need for change (in terms of design or method of working). The range of specialist services now available is such that various teams may be on site at different times and often employed by different subcontractors. By the nature of the works, a lot of complex data is gathered often over a large area during a ground investigation.

Under CDM 2015 (HSE, 2015), the Principal Designer is responsible for the production of the pre-construction information pack (PCIP). This information is important to help all parties understand site constraints, potential site hazards and allow effective planning of the delivery stage. If this information is available in a user-friendly format that can be updated as work progresses, then it will ensure more efficient planning with risks mitigated. The adoption of geographical information systems (GIS) is becoming more widespread in achieving this aim.

Project controls including effective procedures, site briefings and clear lines of communication are important to ensure all parties are informed and standards maintained. Lines of communication need to extend beyond site boundaries and include chain of custody with respect to samples (especially potentially hazardous samples) selected for laboratory testing.

Shared office space, integrated logging facilities, collaborative workshops and daily briefings at a site level all help to encourage working towards shared goals. Social value initiatives, stand-downs or simply celebrating project milestones can also encourage aligned values and can bring disparate businesses together to deliver as a single coherent team.

Shared experience and learning also allows training and development of staff, continual improvement of systems, procedures and working methods that yield safer delivery of works. ISO 44001 (BSI, 2017) provides a roadmap for establishing and managing collaborative relationships for the benefit of all parties.

4.3. Selecting a ground investigation contractor

When selecting a ground investigation contractor, private procurement Clients should satisfy themselves that

- bids are sustainable, which includes understanding the costs over the lifetime of the project
- bids meet quality standards
- risks and uncertainties have been identified and adequately priced.

They should base tenders on best value and past performance rather than cost alone and consider incorporating performance incentives. The use of longer-term frameworks over single-project tender processes could be beneficial but, if used, the frameworks should be subject to periodic market testing to ensure they deliver value for money.

Where Clients enter into collaborative contracts they should follow through with the agreement throughout the lifetime of the project and not just at its inception, even when it becomes commercially challenging. They should carefully choose independent specialists (Designers/consultants) to ensure they have the right skills and experience to provide robust assurance that projects are being delivered efficiently and to cost, quality and time.

Clients should carry out due diligence to gain assurance that Contractors understand their own costs and have robust and well-controlled management information infrastructures in place.

Additionally, for projects that carry a high level of risk, they should consider independent assurance of Contractor competence. This can be achieved by appointing an independent ground investigation specialist, third party accreditation and/or completion of a prequalification assessment process. Clients should recognise the benefits of driving the use of innovation in processes and technology. They should also communicate the importance of equity and diversity as part of the tender and seek to understand the diversity policies of firms that are bidding for contracts.

4.4. Early contractor involvement

Early contractor involvement (ECI) is the process by which a Contractor is engaged by the Client to take part in the planning stage of a ground investigation. For ground investigation projects, the experience and skills of the ground investigation contractor are surprisingly often overlooked in procurement. Adopting an ECI phase can create huge benefits for the Client and add invaluable insight to the execution of an effective ground investigation; it should be considered for all projects either in a formal or informal capacity, as discussed in more detail below.

The Latham Report (Latham, 1994) was commissioned by the UK government and members of the construction industry. The goal was to review procurement and contractual engagement within the construction industry, which was ailing after the recession of the early 1990s. Although the term 'early contractor involvement' was not cited within the report, Latham identified the industry as adversarial, ineffective, fragmented, incapable of delivering for its Clients and lacking respect for its employees. Teamwork between supplier and Client was recommended to encourage an atmosphere of openness, cooperation, trust, honesty, commitment and mutual understanding within the whole project team to produce increased efficiencies and cost savings.

The Government Construction Strategy (Cabinet Office, 2011) was published with the aim of producing efficiency and cost savings within public sector construction projects. The report stated that

> the full engagement of an integrated team – with Designers, other professional consultants and constructors offering an integrated proposition, and with key trade Contractors and manufacturers involved in developing the design (that is, early supply chain involvement)

was key to the strategic goal.

Important Note

In 2015, an Early Contractor Involvement Clause was added to the NEC3 suite of contracts. The guidance note stated, 'Early Contractor involvement (ECI), where a Contractor takes part in the design development and construction planning stage of a project, is now widely recognised as beneficial.'

The NEC4 Engineering and Construction Contract (ECC) issued in 2017 (NEC, 2017) has gone further still. Secondary Option X22 for early Contractor involvement is included. The key points of Option X22 are summarised by the NEC as follows.

- NEC4 ECC Option X22 on early contractor involvement offers Clients a two-stage approach to procurement.
- The latest amendments provide greater flexibility to agree changes during stage one before moving on to stage two construction.
- Successful early contractor involvement relies on collaboration and adherence to the entire conditions of contract.

Seeking the engagement of a ground investigation contractor in the design of a smaller, more complex investigation can also provide significant benefits. This can include, but is not limited to, the correct selection of ground investigation plant and apparatus, promotion of safer systems of work and innovative ground investigation techniques.

During the procurement process, the onus is on the Designer to promote the benefits and efficiency savings of ECI to the Client. This may involve paying a small consultancy fee to the ground investigation contractor for input or a full ECI procurement as detailed below.

Two routes are proposed for appointing a ground investigation contractor to a project where ECI is proposed.

▨ A two-stage tender process: an unrefined version of the project is put out to competitive tender to select a preferred Contractor. The appointed Contractor is then contractually engaged within an ECI contract and works with the Designer and Client to refine the specification. Once final design has been agreed, the ground investigation Contractor will then be contractually engaged to undertake the ground investigation.

▨ A call-off contact within an existing framework: the Client may choose to procure a framework ground investigation contractor directly to engage within an ECI ahead of a ground investigation commencing.

ECI provides the following beneficial facets to a ground investigation project, such as

▨ detailed knowledge and experience of the expected ground conditions, helping to inform the scope of the ground investigation

▨ input into the correct and most appropriate ground investigation techniques and the most efficient and cost-effective way to deliver a ground investigation

▨ assisting with landowners and stakeholder liaison, where needed

▨ highlighting site-related risks to reduce risk of unexpected cost increases

▨ supporting the Client and Designer in the design and costing of any necessary enabling works

▨ design and costing of site compounds and other infrastructure required for ground investigation delivery

▨ providing innovation proportionate to the scale and complexity of the ground investigation

▨ providing guidance on the use of specialist ground investigation techniques and Contractors

▨ production of an accurate ground investigation project programme

▨ development of the ground investigation project team, lines of communication and ways of working

▨ building a better team ethic, improving interorganisational relationships and enabling more effective collaboration

▨ developing and improving project understanding of health, safety and quality requirements

▨ assisting the Client with the development of ground investigation contract documentation if needed, where a Client organisation lacks internal resources to perform this function.

4.5.   Types of procurement approaches
4.5.1   Public procurement

Public procurement is the process through which supplies, services and works are acquired on behalf of the public and nonprofit sectors. The Crown Commercial Service (CCS) is the government organisation responsible for setting the legal framework for some public procurement, incorporating relevant directives and regulations, and for developing and implementing the associated policies and guidance (CCS, 2023). It should be noted that there are other organisations who receive direct public procurement instructions from central government and in particular the Cabinet Office.

At the core of public procurement policy is a requirement for value, defined by the CCS (2023) as

the best mix of quality and effectiveness for the least outlay over the period of use of the goods or services being bought.

All central government procurements set thresholds and are competitively tendered following the Public Contract Regulations 2015 (PCR) (HMG, 2015). The PCR promotes free and open competition, ensuring value for money based on a combination of quality and cost criteria to award the most economically advantageous tender (MEAT). The public sector procurement process is essential in ensuring that taxpayer money is used efficiently. Unless there are compelling reasons, value is achieved through open, fair and transparent competition, and ensured by the legal framework that governs public procurement. That legal framework is built on international obligations that are implemented through national regulations or legislation.

In addition to the PCR, procurement policy notes (PPN) set out information and guidance for public sector bodies who wish to procure works, goods or services. They are aimed at giving transparency to the procurement process and ensure that tendering practices are ethical. The notes are approved by the government and are mandatory for government departments to implement. A list of PPN updates can be found on the government website, providing guidance on best practice and conditions for public sector procurement.

The *Construction Playbook* (Cabinet Office, 2022) is focused on getting projects and programmes right from the start, regardless of what is being bought, and sets out important policies and guidance on how to procure and manage public works projects and programmes.

The PCR, procurement policy notes and Construction Playbook combined, provide a set of legislation, guidance and best practice for public sector bodies tendering construction-related contracts.

Although the UK left the European Union in 2020, the UK government implemented three EU procurement directives, namely

- Public Sector: Directive 2014/24/EU
- Concessions: Directive 2014/23/EU
- Utilities: Directive 2014/25/EU.

All handbooks, guidance documents and training materials for the above can be found on the government website.

Note: the Procurement Regulations have been reformed and the revised regulations are due to come into force in 2025. For further information on the Procurement Regulations 2024, please see the government link to Transforming Public Procurement here: https://www.gov.uk/government/collections/transforming-public-procurement

The Cabinet Office have produced a series of commercial playbooks, including the UK *Construction Playbook* (Cabinet Office, 2022), which includes a model clause relating to subsurface data sharing. This clause is provided for insertion into contracts where subsurface data will be collected, including ground investigation contracts, to require data to be shared with the National Geoscience Data Centre (the British Geological Survey).

Public procurement commonly adopts a formal prequalification process to assess the competence and capability of ground investigation Contractors. This may be in the form of a questionnaire, often referred to as a prequalification questionnaire (PQQ) or interview and/or presentations.

4.5.2   Private procurement

While public and private sector procurement approaches may appear similar, there are subtle differences. Public sector organisations operate in the interest of the public, so are guided by principles that benefit the needs of society, rather than by a need to be profitable. Conversely, the private sector is profit driven and may involve transactions that are private, so not subject to the principles of equality of treatment, transparency and mutual recognition and proportionality that exist in the public procurement approach.

The absence of the legislative requirements and governance makes private procurement simpler for a Client organisation, although the approach may be less fair, open or transparent, and so potentially be less competitive for suppliers. Private sector Client organisations are freer to choose what is best for them.

Private procurement has fewer regulations; however, those that are still upheld include law for monitoring such things as equality and bribery.

A key difference in private procurement is finance and budgets; private sector businesses often have fewer legislative constraints and governance and may have greater, unrestricted purchasing power. This can result in greater flexibility during the procurement process.

Private procurement is often more dynamic than public procurement and allows for quicker decisions to be made, resulting in a time-efficient process with greater fluidity of transactions.

The reason for this is that private sector organisations are financed using private money through investors, shareholders, loans and other forms of finance. While private organisations often still have budgets set for aspects of their project, they are far more flexible and dynamic, and money can be easily moved between different project phases in response to changing project requirements.

The private sector procurement process is primarily driven by established relationships built on trust, reliability and competency. The tender process is cooperative and collaborative, with many aspects of the procurement process open to negotiation, such as tender deadlines and project milestones.

Private sector procurement does not usually have a prequalification process and suppliers are instead directly issued an invitation to tender (ITT). In general, there is no contractual relationship arising from the ITT. It is more often implied that a tendering contract relationship exists whereby certain aspects of the ITT process are binding and enforceable – for example, provisions dealing with confidentiality, nondisclosure, withdrawal, cancellation and liabilities. These aspects are often defined in the exclusion clauses.

An ITT should contain express statements on how the bid process will be managed and awarded. Within the ITT there is often a requirement for the tenderer to provide a warranty, performance bond or parent company guarantee (see Appendix C for descriptions of these terms) and disclose full financial accounts such that the Client's liabilities are protected and the risk passed through the supply chain. Very rarely does the Client take any risk and often demands multi-million-pound professional indemnity insurance with cover for *each and every* claim.

In private sector procurement it is accepted that the tenderer takes full risk for their tender costs. However, where the tenderer has been requested to carry out initial preparatory works, to be involved at early stages of procurement (early contractor involvement (ECI)) or to contribute to the design of the project specification, they may be entitled to recover the relevant expenses incurred under a separate billing transaction or within the body of the project's ITT.

4.5.3   Framework contracts

Framework contracts, as the name implies, involve an overarching agreement between a buyer and one or more suppliers, allowing for flexible procurement of services over a defined term or across discrete packages of work. Under the Public Contract Regulations 2015 Regulation 33(2) (HMG, 2015) a framework agreement is defined as

> an agreement between one or more contracting authorities and one or more economic operators, the purpose of which is to establish the terms governing contracts to be awarded during a given period, in particular with regard to price and, where appropriate, the quantity envisaged.

Frameworks can provide flexibility to the Client through the option to operate a call-off arrangement, secondary competitions for discrete packages, lots or projects which are offered to one or more framework Contractors. They can be of varying scale from modest short-term single-supplier arrangements to major national public sector long-term multi-supplier contracts. The larger agreements may be divided into packages by service type or region. A framework can provide the advantage of reducing the scale of individual work packages, encouraging greater market participation, especially by smaller suppliers who may otherwise be unable to address the overarching project or programme requirements.

A variant of the framework introduced in the UK public sector referred to as 'dynamic purchasing systems' is essentially an approved list of suppliers but has less certainty than a framework, in that terms are not usually agreed in detail and suppliers may join or new services can be added at any point.

An allocation process for a framework will be set out in the contract documents. Processes may differ but initial allocation of work may be based on a supplier's initial tender score for a specific package of work. Thereafter allocation may be on a rotational basis, awarding to each tenderer in turn, if they are compliant and pass a set of defined criteria. This helps to ensure the allocation process is fair and that work is spread evenly across the relevant package of work. Suppliers can be scored on their performance and any supplier not achieving a set pass threshold would exclude itself from allocations until such time it developed and implemented an improvement plan and performed above the accepted threshold.

If a supplier declines an allocation, it can be offered to the next eligible supplier in the relevant package of work. If a direct allocation is refused by all suppliers in the relevant package of work, a secondary competition should be run involving all eligible suppliers who are of course free to decline the tender opportunity.

A secondary competition may be run if it is felt to provide most value to the Client or if the scheme was not awarded by way of an allocation. Secondary competition can help support efficiency, innovation and quality, with the assessment process being able to focus on those areas that are important for a project and select the bid that can supply the best balance of quality against cost.

It is crucial that government, local authorities and arm's-length bodies structure secondary competitions in the most beneficial way to achieve a quality project at the best value for money. Examples of current frameworks include the National Highways GI Framework and the Coal Authority Framework.

4.6.   Assessment criteria

Public Clients have clear obligations in evaluating and considering tenders; however, unless it has committed to do so, a private Client is not required to accept the lowest bid; although it is recommended that ITT documentation makes this point clear, it is generally accepted that it is unnecessary for the Client to include an assessment criterion or a statement relating to award criteria. Private sector procurement allows Client discretion or preference to be applied when it comes to choosing the winning bid. Where an assessment criteria or statement relating to award criteria is included in the ITT, it is usually accompanied by a statement that is nonbinding such that the Client retains a discretionary right to award the tender as it sees fit, or not even award at all. Notwithstanding that, the employer has a duty to act fairly and in good faith to consider all compliant bids.

Public sector assessment criteria have to follow a more rigorous and prescriptive process, as detailed within the guidance produced by the UK government.

4.7.   Sustainability and carbon reduction
4.7.1   Introduction

Sustainability in the ground investigation sector can be related to three key factors, namely environmental, economic and social. The procurement of a truly sustainable project will embrace all three of these factors and promote high standards in each. Procurers should examine the 17 United Nations' Sustainable Development Goals (SDGs) (UNDESA, 2024) and use these to further inform their priorities in procurement. Although procurement is often very focused on economic factors, and consequently tends to influence SDG 8 Decent Work and Economic Growth more than other SDGs, many others can in some way be influenced by procurement processes, some more easily than others. Deamer et al. (2021) indicate that the ten SDGs shown in Figure 4.2 are most influenced by procurement functions.

The sustainable procurement of ground investigations involves every level of the project supply chain, from Client to subcontractor, with opportunities for betterment through dialogue, design and implementation. The Contractor selection process can allow identification of organisations aligned with sustainable goals at supply chain management level (ISO 14001 certification (ISO, 2015) etc.), while selective targeting of sustainability goals during the project-specific procurement process can embed strategic objectives within individual project delivery plans.

Clients should decide if they want to differentiate between supplier requirements at a project and company level. Ideally Clients should be moving towards embracing a completely sustainable supply chain and consider at the design stage whether alternatives can be identified to achieve a more sustainable ground investigation.

Figure 4.2 Sustainable development goals influenced by procurement functions (courtesy of the United Nations https://www.un.org/sustainabledevelopment. The content of this publication has not been approved by the United Nations and does not reflect the views of the United Nations, its officials or member states)

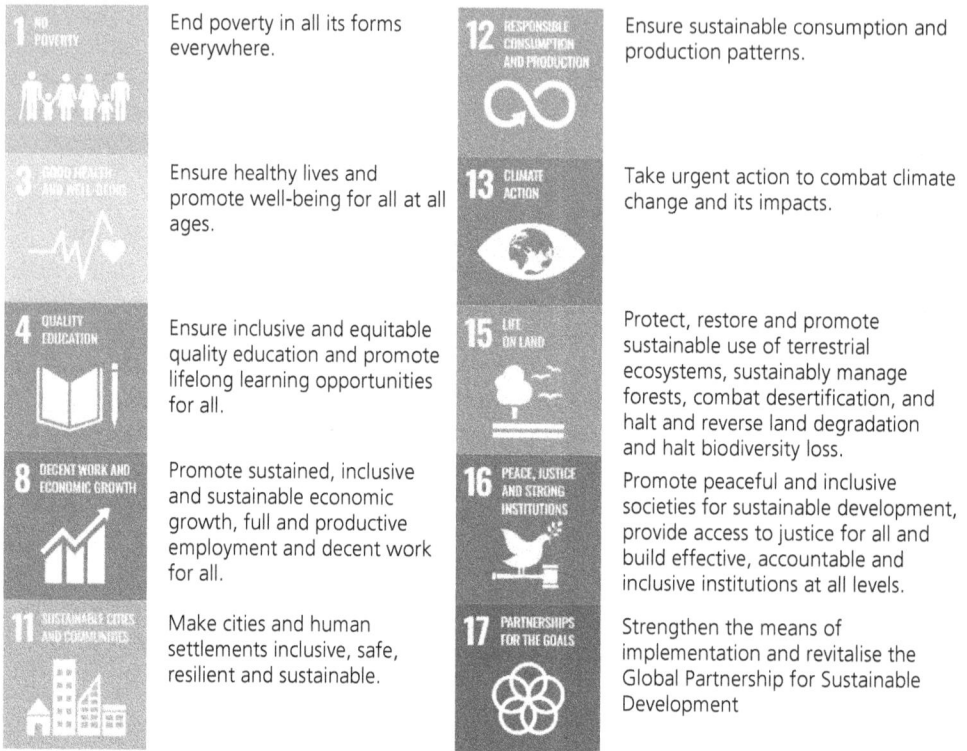

1 NO POVERTY	End poverty in all its forms everywhere.	12 RESPONSIBLE CONSUMPTION AND PRODUCTION	Ensure sustainable consumption and production patterns.
3 GOOD HEALTH AND WELL-BEING	Ensure healthy lives and promote well-being for all at all ages.	13 CLIMATE ACTION	Take urgent action to combat climate change and its impacts.
4 QUALITY EDUCATION	Ensure inclusive and equitable quality education and promote lifelong learning opportunities for all.	15 LIFE ON LAND	Protect, restore and promote sustainable use of terrestrial ecosystems, sustainably manage forests, combat desertification, and halt and reverse land degradation and halt biodiversity loss.
8 DECENT WORK AND ECONOMIC GROWTH	Promote sustained, inclusive and sustainable economic growth, full and productive employment and decent work for all.	16 PEACE, JUSTICE AND STRONG INSTITUTIONS	Promote peaceful and inclusive societies for sustainable development, provide access to justice for all and build effective, accountable and inclusive institutions at all levels.
11 SUSTAINABLE CITIES AND COMMUNITIES	Make cities and human settlements inclusive, safe, resilient and sustainable.	17 PARTNERSHIPS FOR THE GOALS	Strengthen the means of implementation and revitalise the Global Partnership for Sustainable Development

4.7.2   Environmental sustainability

Construction activities, including ground investigations, are an acknowledged major contributor to global carbon emissions, primarily by way of direct routes including plant and vehicle movement, as well as through offsite activities and embodied carbon related to materials. As the key product of a ground investigation is data, rather than a physical object, simple substitution of concrete and steel may not be a viable option in reducing carbon embodied within the product. The techniques used to obtain the data should be assessed and the industry should consider alternatives to reduce carbon. In turn, procurement programmes must encourage and incentivise low-carbon approaches and related initiatives to support delivery partners in this sphere.

More environmentally sustainable ground investigation begins with intent and can be embedded throughout the process by way of design requirements, collaborative approaches and an understanding that cost and benefit are not measured in currency alone. Strategic planning of ground investigation scopes, considering diverse data sources, innovative techniques and adaptive approaches to testing and sampling can reduce embodied carbon in ways as listed below.

- Assessment of prior data for validity and relevance which may decrease the quantity of intrusive work required. Consider undertaking validation investigation for historic data rather than wholly overprinting it with new investigation locations and techniques.

- Scheduling of intrusive techniques alongside nonintrusive methods to build a lower carbon ground model (paired use of geophysics and intrusive analysis).
- Adaptive approach to sampling and testing – can the number of samples taken and, therefore, the number of containers used be decreased? Or can increasing sampling in some locations balance a change to a lower carbon intensity technique in another? Should all project samples be shipped to the laboratory? Or can they be selected on site and cut carbon embodied in transport?
- Plan welfare facilities to maximise the use of renewable power sources and/or low carbon equipment. Can the Client make adequate space available to facilitate solar power, battery storage and so on?
- Manage the environmental impact of fieldwork through surface damage, silting and watercourse impacts, decreasing the effect of works on biodiversity, or provide appropriate reinstatement to maintain or improve biodiversity within a site.
- Promotion of carbon reduction technology including anti-idling shutoff, alternative fuel or power supplies and even the provision of site access to minimise plant and vehicle transit distances can all have a significant impact on the reduction of carbon expenditure during fieldwork.

4.7.3 Economic sustainability

The economic viability of ground investigation Contractors and their own supply chains is critical to both the retention of knowledge and capability within the sector, and the support of investment in equipment, research and development. A suitable level of due diligence by a Client prior to appointment of a Contractor can support the awarding of work to businesses who operate in a sustainable fashion, with risk managed, any financing arrangements secured by way of socially responsible bodies, and with payment practices that support the sustainability of their own supply chain. Clients should be conscious that their own payment practices can have significant bearing in the sustainability of their suppliers. Fundamentally, the stability of a Contractor providing design data is vital to support interrogation of that data. Where necessary, it provides reliable access to recourse should there be issues with that information in the future, potentially leading to claims. The economic viability of a business has a direct impact on the wellbeing of staff and supply chain members through continuity of employment.

Vetting of suppliers to assess their viability is a key approach to sustainable procurement. A variety of methods can be employed, based on the scale of the Client or project. These include wide-ranging external assessments such as Achilles or Constructionline, more financially focused appraisal such as Dun and Bradstreet ratings, or an internally completed assessment of financial ratios.

4.7.4 Social sustainability

Embedding social value in ground investigation can be seen as both a business advantage and a timescale challenge from a Contractor's perspective. Maximising the use of local subcontractors not only retains project value in the local community but also has the potential to decrease a project's carbon footprint through reduced mobilisation distances and workforce travel needs. In contrast, building relationships with community organisations and delivering support to the education sector can be challenging on short-timescale projects, and is, therefore, an area where collaboration is critical. The use of a localised workforce can also bring staff wellbeing benefits by way of reduced travel time and improved rest periods.

Socially proactive sourcing, placing emphasis on the use of local suppliers and workforce can be achieved through a variety of means.

- Tender structure: the use of localised frameworks or regional work packages provides scope for the development and positive exploitation of local supplier networks.
- Tender assessment criteria: including local spend proportion within tender scoring criteria adds an up-front requirement to investigate and deliver opportunities.
- Project incentives/expectations: contractual obligations based on the above and/or paired with social engagement activities can focus project delivery on maximising local spend and localised value retention.

Social engagement and support can prove challenging to deliver within the very short timescales of ground investigation projects. However, through relationship building and collaborative working, even the shortest of onsite durations may be able to yield social benefit. While many public sector contracts utilise the index of quantified social value contributions 'Social Value Model Quick Reference Table', it remains a challenge for a Contractor alone to embed within a community on a small project. However, if the Client and their consultants are willing to embrace a shared responsibility to deliver social benefits throughout a scheme, this can become a shared success. Consideration of any planning constraints, and even the interested parties in a planning review, may yield opportunities to deliver early-stage benefits to a local community. Table 4.1 provides examples of social engagement for different size projects.

Table 4.2 provides a checklist for the evolution of sustainable procurement.

Table 4.1 Examples of social engagement for different size projects

Small to medium	Large and/or complex
Donating core samples to local geology departments	Site visits
Work experience	Work placements
Talks to schools and unemployed/career engagement workshops	Community engagement events – for example, in the evenings – educational
Social enterprise or community group donations	Employment/employability support including CV workshops and mock interviews

Table 4.2 Checklist for the evolution of effective sustainable procurement (continued on next page)

Client procurement checklist

☐ Have all significant risks (commercial, technical, health, safety and environmental) been
  ☐ identified?
  ☐ understood?
  ☐ fairly allocated?
  ☐ appropriately costed?

☐ Has a sustainable margin been included (including contingency to cope with risks that might occur)?

Table 4.2 Continued

Client procurement checklist

☐ Does the Contractor have all of the necessary skills to deliver the proposed project, including skills and ability to meet unfamiliar technical necessities/complexity and sustainability criteria?

☐ Has a skill and resource assessment at the prebidding stage been carried out to ensure there is the capacity and capability to deliver the required level of quality and key sustainability objectives?

☐ Do the decision makers actively refer to and consider lessons learnt historically when determining whether to approve the bid?

☐ Have discussions been held with Contractors or a formal early contract involvement process been started?

☐ If the Contractor is relying on its supply chain to deliver additional skills not available in house or other significant services in connection with the contract
  ☐ Is there sufficient evidence available around previously demonstrated capability, access to necessary resources and financial health?
  ☐ Are key supply chain partners actually in place at the time of the bid's submission and do they have the proper level of involvement in the bid process?
  ☐ Are subcontractors' prices locked in at the bid stage?
  ☐ Have the subcontractors and supply chain products been incorporated into the sustainability deliverables and goals in the supply chain management process?

☐ Has the proposed form of contract been reviewed by persons with sufficient expert knowledge to highlight key risk areas and drafting advice designed to provide adequate protection in the event of a dispute?

☐ Have timescales for the bidding process and actual works been stress tested against current resource availability and market conditions?

☐ Does the contract provide a common understanding and certainty to all parties?
  ☐ Will it deliver the data required to fulfil the purpose of the investigation?
  ☐ Will it meet the key sustainability objectives?
  ☐ Is it clear what is covered and not covered in the original contract?
  ☐ Who will bear the cost of any changes?

For larger, higher risk or more complex projects the following additional checks would be prudent.

▥ Does the Contractor involve a suitably qualified panel of decision makers to evaluate whether the proposed price/return is adequate considering the risks accepted – and are the go/no-go reasons and relevant supporting evidence retained for future reference and for use in future bids?

▥ Are the right parameters in place to ensure decisions are right for the wider business? For example, does the management team have final sign off? Has the board been informed about the risks and costs from the outset to ensure the bid is sustainable?

▥ Do the decision makers actively refer to and consider lessons learnt historically when determining whether to approve the bid?

▥ Have anticipated costs been stress tested (looking at potential adverse circumstances and scenarios) to see how the project would be affected by different external factors (e.g. supply chain inflation, skills shortage, lack of resources, land access issues)?

> ### CASE STUDY
>
> ## Early contractor involvement
>
> | | |
> |---|---|
> | *Project size* | Medium |
> | *Informed Client* | Yes |
> | *Geotechnical specialist appointed* | No |
> | *Conditions of contract* | NEC3 |
>
> ### Background
>
> Ground investigation works were required to assess an unstable slope which was creating trafficability and safety issues for a road. Assessment of the ground conditions and data was required to allow long-term engineering solutions to be determined while still allowing traffic to use the road.
>
> ### Challenges
>
> - The site was adjacent to a busy trunk road.
> - Traffic management had already been employed to restrict traffic to single lane controlled by traffic lights.
> - Working was close to live traffic.
> - Limited working areas were available.
> - Numerous stakeholders were involved.
>
> ### Positive actions
>
> - Quick identification and appointment of the ground investigation contractor occurred.
> - Open discussions between the Client and the ground investigation contractor were held.
> - An early walkover between the Client and the ground investigation contractor was held.
> - Risks, especially related to health and safety, were identified quickly and control measures proposed and agreed.
> - ECI improved the planning of the works, allowed quick decisions to be made and ensured the commercial delivery of the project.
>
> ### Lessons learnt
>
> - ECI creates a positive collaborative environment without the pressures which result from site works.
> - Involving stakeholders prior to commencing on site and allowing the ground investigation contractor to be involved obtains positive engagement.
> - Allowing the ground investigation contractor to identify challenges and potential delays prior to commencing reduces impact and disruption.
> - A clear understanding of the objectives at the outset allows for more effective planning.
> - Preplanning during ECI allows for a quicker execution when on site.
> - Identification of risk, risk allocation and control measures during ECI reduces Client exposure and additional costs.

> ## CASE STUDY
>
> ## Socio-economic and environmental sustainability
>
> | | |
> |---|---|
> | *Project size* | Medium |
> | *Informed Client* | Yes |
> | *Geotechnical specialist appointed* | Yes |
> | *Conditions of contract* | NEC3 |
>
> ### Background
>
> A regionally significant ground investigation was planned with the intention of unlocking socio-economic development potential through infrastructure development. The ground investigation was intended to improve regional and local connectivity and support development opportunities.
>
> ### Challenges
>
> - Maximising social and environmental impact of the project should be instigated at an early stage as this helps to set a precedence to be followed throughout the construction phase, embedding social value and societal sustainability into the scheme.
> - Executing works with both technical and sustainability goals in mind, without compromising on either, can be difficult.
>
> ### Positive actions
>
> - The Contractor planned to minimise road transport mileage throughout the project and the decrease of empty vehicle miles.
> - The priorities were cascaded through the supply chain side of the scheme to maximise impact.
> - The priorities were reinforced at internal procurement and logistics levels in both onsite and offsite activities.
> - The Client was able to influence social value activities and provide support to Contractors and the community alike to align capabilities and needs.
> - The Client identified and agreed key targets to focus on: education and careers; social cohesion; and social history and culture.
>
> ### Lessons learnt
>
> - Through a combination of onsite/offsite planning and collaborative working practices the stakeholders embedded a carbon reduction culture within the project, facilitating practice development within their wider business.
> - In supporting outreach projects through both time and financial donations the collaboration between Client and Contractor promoted social sustainability in career visibility, social cohesion and cultural preservation.
> - HGV transport planning, maximising use of 'backloads' to avoid empty vehicle mileage requires planning and flexibility but can cut emissions effectively by up to 50% per vehicle load.
> - Education and careers were addressed by way of multiple school careers events with a diverse range of staff attendees. A school visit to the regional office and workshop were hosted to showcase the wide range of careers contributing to the geotechnical sector.
> - Social cohesion was addressed by financial support to a local mental health charity.
> - Social history and culture were supported through donations to a local historic transport preservation trust.

REFERENCES

BSI (2017) ISO 44001: Collaborative Business Relationship Management Systems – Requirements and Framework. BSI, London, UK

Cabinet Office (2011) *Government Construction Strategy*. The Stationery Office, London, UK. https://assets.publishing.service.gov.uk/media/5a78ce8eed915d07d35b2933/Government-Construction-Strategy_0.pdf (accessed 09/10/2023).

Cabinet Office (2022) *The Construction Playbook*. The Stationery Office, London, UK. https://www.gov.uk/government/publications/the-construction-playbook (accessed 09/10/2023).

CCS (Crown Commercial Service) (2023) Public Procurement Policy. https://www.gov.uk/guidance/public-sector-procurement-policy (accessed 09/10/2023).

CIPS (Chartered Institute of Procurement and Supply) (2023) Procurement Fundamentals. https://www.cips.org/intelligence-hub/procurement-fundamentals (accessed 09/10/2023).

Deamer L, Lee J, Mulheron M and De Waele J (2021) Building sustainability impacts from the bottom up: Identifying sustainability impacts throughout a geotechnical company. *Sustainability* **13(21)**: 11981. https://doi.org/10.3390/su132111981 (accessed 11/04/2024).

HMG (Her Majesty's Government) (2015) Public Contracts Regulations 2015. The Stationery Office, London, UK, Statutory Instrument 2015 No. 102. Open Government Licence v3(0).

HSE (Health and Safety Executive) (2015) Construction (Design and Management) Regulations. The Stationery Office, London, UK.

ISO (International Organization for Standardization) (2015) ISO 14001: Environmental Management Systems. ISO, Geneva, Switzerland.

Latham M (1994) *Constructing the Team: Joint Review of Procurement and Contractual Arrangements in the United Kingdom Construction Industry*. The Stationery Office, London, UK.

NEC (New Engineering Contract) (2017) *NEC Contracts*. NEC, London, UK. https://www.neccontract.com/ (accessed 09/10/2024).

UNDESA (United Nations Department of Economic and Social Affairs) (2024) Sustainable Development: the 17 goals. https://sdgs.un.org/goals (accessed 09/10/2024).

FURTHER READING

https://www.legislation.gov.uk/uksi/2015/102/part/2/chapter/1/made

https://www.gov.uk/government/collections/procurement-policy-notes

https://assets.publishing.service.gov.uk/media/6312222dc90e075880923330/14.116_CO_Construction_Playbook_Web.pdf

https://assets.publishing.service.gov.uk/media/61b9cb41e90e07043e8ff5cc/Constructing_The_Gold_Standard_Final.pdf

The Public Contracts (Scotland) Regulations (2015). Scottish Statutory Instruments No. 446. Open Government Licence v3.0.

EFFC Carbon Calculator. https://www.effc.org/how-we-operate/eco%E2%82%82-foundations/

https://assets.publishing.service.gov.uk/media/5fc8b804d3bf7f7f53e5a503/Guide-to-using-the-Social-Value-Model-Edn-1.1-3-Dec-20.pdf

https://assets.publishing.service.gov.uk/media/5fc8b8248fa8f547499d79b7/Social-Value-Model-Quick-Reference-Table-Edn-1.1-3-Dec-20.pdf

World Trade Organisation (2012) Agreement on Government Procurement. https://www.wto.org/english/tratop_e/gproc_e/gp_gpa_e.htm (accessed 09/10/2023).

emerald
PUBLISHING

ice
Publishing

Association of Geotechnical and Geoenvironmental Specialists
ISBN 978-1-83662-029-7
https://doi.org/10.1108/978-1-83662-026-620251005
Emerald Publishing Limited: All rights reserved

Chapter 5
Establishing effective contracts

5.1.   Introduction

Contracts are binding agreements with specific terms between two or more parties in which there is a promise to do something in return for a valuable benefit (usually payment). Contracts can be in writing or verbal. It is vital that an effective and fair form of ground investigation contract is established to enable the best acquisition of ground data to be obtained in a cost-effective way for a Client. Significant future cost savings can be made on a project and numerous risks avoided by the provision of robust and reliable data.

It is important to understand the differences between a ground investigation contract and standard 'building' contracts. Although various monitoring installations may be regarded as the products of 'construction', the main aim of the 'works' will usually be to inform the Designer of the ground conditions. There are aspects in a ground investigation contract that result in crossing the boundaries between the undertaking of physical 'works' (sinking boreholes, excavating pits, specialist nonintrusive methods and undertaking testing etc.) and the provision of advice (logs, grouping of strata, characteristic values etc.).

Ideally, the form of contract for a ground investigation should

- have the confidence of Clients/Employers, Designers and Contractors
- be able to effectively deal with changes
- allow for the most efficient and safe acquisition of data
- be applicable to 'works' within an office, site and laboratory environment
- ensure that the unique risks of ground investigation are fairly attributed between parties, with responsibilities clearly defined
- reduce disputes
- work alongside the *UK Specification* (AGS, 2022)
- be able to deal with innovation within the ground investigation industry.

5.2.   Applicable forms of contract
5.2.1   Introduction

Ground investigations are carried out under various forms of contracts and subcontracts, generally reflecting the experience and opinions of the Client, together with their type of business. The most applicable forms of contract to ground investigation are

- Infrastructure Conditions of Contract, Ground Investigation Version
- NEC4 Engineering and Construction Contract (ECC) and Engineering and Construction Short Contract (ECSC).
- Professional services contracts such as the NEC4, Professional Services Contract (PSC) and Professional Services Short Contract (PSSC).

Appendix C provides detailed information on several specific contract clauses and provisions which may form parts of a ground investigation contract.

5.2.2   Infrastructure Conditions of Contract, Ground Investigation Version

The Infrastructure Conditions of Contract, Ground Investigation Version (ICC Conditions) is the only ground investigation-specific form of contract published in the UK. It is based on the Institution of Civil Engineers (ICE) Conditions of Contract, Ground Investigation Version, second edition, 2003. This, in turn, was based on the ICE Conditions of Contract, fifth edition, 1973.

The Association for Consultancy and Engineering (ACE) and the Civil Engineering Contractors Association (CECA) published the ICC Conditions in 2011 following the decision by the ICE's publishers to not update the second edition. This format relies on claims made by the Contractor for extensions of time and additional payments, although there are some options for change of scope.

The ICC form is ideally suited to ground investigations as it includes ground investigation specific clauses relating to items such as

- unavoidable damage
- care of samples and cores
- dealing with fossils, coins, articles of value or antiquity and geological/archaeological interest
- the requirement for the Contractor to investigate alternative measures when the investigation reveals conditions which change the project
- contamination.

Due to the historical format of the ICC Conditions, they work well with the *UK Specification* (AGS, 2022), including

- the designated roles of the Investigation Supervisor
- remeasurement using a bill of quantities format
- incorporation of the CDM Regulations (HSE, 2015).

The format is familiar within the ground investigation community, where most of the issues relating to working with the contract are well known.

Appendix D provides a glossary of terms which may be found in the ICC Conditions of Contract.

5.2.3   NEC4 ECC and ECSC contracts

NEC (2017) produce collaborative, programme-driven contracts, with provisions to deal with early warnings, change events and situations that change the scope of works. Although there is no specific NEC ground investigation contract, various NEC forms are routinely adopted for ground investigations. These are construction contracts that can be used for ground investigation works.

The NEC produced guidance on using NEC contracts for ground investigations in June 2023 (https://www.neccontract.com/news/new-guidance-on-using-nec-for-ground-investigations). This guidance demonstrates how ground investigation contract works can be procured using NEC forms and gives advice on how to implement this, dealing with the specific requirements of ground investigation contracts. Because of the nature of the physical work, one of the 'works' contracts in the NEC family will normally be appropriate, either the Engineering and Construction Contract (ECC) or the Engineering and Construction Short Contract (ECSC). While a Professional Service Contract (PSC) could be appropriate for desk studies, guidance and reporting, it is generally not suitable for site works where there is no significant element of Contractor design. NEC advises that simple, straightforward ground investigations could use the ECSC, with the ECC being more suitable for larger scale, more complex and higher risk projects, where an 'Option B' contract allows for a bill of quantities such as those contained in the ICE specification. The ECSC generally provides less protection for the Contractor, the programme is not so important, but there is an easier process for dealing with compensation events. The NEC also produce guidance on the comparison between the ECSC and ECC forms of contract (https://www.neccontract.com/news/comparison-of-the-nec4-ecsc-with-the-nec4-ecc).

The ECSC includes in clause 63.1

> For a compensation event which only affects the quantities of work shown in the Price List, the change to the Prices is assessed by multiplying the changed quantities of work by the appropriate rates in the Price List.

Other compensation events are assessed, as is standard for NEC contracts based on the change in (real) defined cost plus fee caused by the compensation event (clause 63.2). The intent is to make the assessment of compensation events that are simple changes in quantities – for example, numbers of boreholes or trail pits – easier and quicker to agree.

If the Contractor is not happy with the rates it has for the items affected, it will try to argue that the event did not 'only affect the quantities of work'. Clause 63.3 also allows the Client and Contractor to agree to use rates or lump sums (from anywhere). The intent of ECSC Clause 63.1 is missing in the ECC.

If that was required a possible additional condition would be to add as Clause 63.0

> For a compensation event which only affects the quantities of work shown in the Bill of Quantities, the change to the Prices is assessed by multiplying the changed quantities of work by the appropriate rates in the Bill of Quantities.

It should be noted that Option A contracts are not subject to measurement. They have an activity schedule as a basis for payment and are very rarely suitable for ground investigations which are generally priced on a remeasurable bill of quantities basis.

Appendix E provides a glossary of terms which may be found in NEC conditions of contracts.

5.2.4 Professional service(s) contracts

Professional service contracts may follow standard terms such as the NEC4 (Professional Service Contract (PSC)/Professional Service Short Contract (PSSC)), ACE or RIBA formats. Ground investigation contracts that follow a specification provided by a Designer/consultant generally cannot be defined as a 'professional service' as there are physical 'works' involved, the ground investigation contractor is not involved in any design (apart from their own temporary works) and the contract should suitably cover items such as

- access, safety, utility services, permits, weather, physical conditions and liability for third party injuries
- special conditions, including contamination, artesian water, voids, coal seams, sample quality, consistency and reliability of logs, factors affecting test results, the definition of 'damage' (which may be unavoidable due to undertaking the investigation in accordance with the specification)
- work items and materials that are specified and measured by others for the ground investigation contractor to price and perform.

It is, therefore, not advised to use a professional service contract for ground investigation works. The ground investigation contractor should not be referred to as a 'consultant' or 'Designer' as this will add a considerable amount of ambiguity to those responsible for the investigation.

There are exceptions where the Contractor has been given significant design responsibility or where the output could be considered as 'advice', such as in the case of an early contractor involvement (ECI) stage prior to a large or challenging ground investigation. If a professional service contract is used for a ground investigation, suitable additional clauses will be needed to cover the relevant physical works items.

5.3. The use of other forms of contract
5.3.1 Introduction
Although the most applicable forms of contract to ground investigation works are the ICC and NEC forms discussed above, there are several other forms used with different approaches to works and services and commensurate levels of risk to the contracting parties. These are discussed briefly in the following sections.

5.3.2 Civil Engineering Contractors Association (CECA) subcontract
CECA published forms of subcontract for use with the ICE Conditions prior to the withdrawal by the ICE's publishers. CECA then decided to publish subcontracts for use with ICC replacements. These are not often used in civil engineering and infrastructure works, primarily because NEC has become the most popular format. Therefore, the subcontracts for use with ICC are also not often used. NEC forms of subcontract, or ICC Ground Investigation Version would be recommended ahead of CECA as experience with these is limited.

5.3.3 Joint Contracts Tribunal (JCT)
The JCT form of contract is generally considered not suitable for ground investigation work. Outside the main complications around cross-referencing of clauses and legal language, it is silent on ground conditions. The Client's lead is the contract administrator and ground risk is the Contractor's risk. The programme is not a contractual document. In terms of managing variations, these are to be carried out on the Client's written instructions only; verbal instructions must be confirmed in writing within two working days. It is important to note that time and money are dealt with as separate concepts.

JCT produce several collateral warranty templates for setting up these agreements between various parties that can be applicable to ground investigation contracts.

5.3.4 Royal Institute of British Architects (RIBA)
Like JCT forms, RIBA contracts are designed for use in building works for consumer or commercial Clients, with a portfolio covering domestic building contracts, commercial and various professional services contracts. These are primarily aimed at the role of architects and are not generally applicable to ground investigation works.

5.3.5 Fédération Internationale des Ingénieurs Conseils (FIDIC)
FIDIC (Fédération Internationale des Ingénieurs Conseils, or later, the International Federation of Consulting Engineers) contracts are mainly used in international engineering contracts. The Red Book is mainly for civil works (the Green Book being the short contract). The Client's lead is the

engineer and there is an order of hierarchy, under 'default hierarchy' or as specified in the contract. Although these terms may be specified by an overseas Client, they are generally not applicable for use on ground investigations.

5.3.6   Client/Consultant/Main Contractor own terms (bespoke agreements)

These are specifically drafted contracts/consultancy service agreements that suit the needs of the instructing party. Risk with variation process mainly favours the Client. It is prudent to understand condition precedent clauses for any claim for extension of time or costs, measurement and evaluation, variations and adjustment, contract price and payment. It is also important to understand risk, responsibility and accountability prior to commencing work on site, particularly with the data requirements that are the subject of ground investigation works.

5.3.7   Supply contracts (for goods and services)

These normally come in the form of a purchase order (PO). They are a contractual instruction issued by Clients to purchase a service. All POs come referenced to their specific terms and conditions. The terms and conditions are usually generic or 'boiler plate' and will have no bearing to site-based ground investigation works. POs generally have no variation process provisions, no detail as to which party is responsible for accesses, permits, licences, consents, wayleaves and so on. This leads to a high risk of disputes among the contracting parties. Most POs will not have signature sections but come with a response period for acceptance. It should be noted that the contract is automatically binding at the expiry of the response period or at commencement of works on site without a prior challenge or agreement to changes.

5.3.8   Specialist forms of contract

There are several forms of contract, such as the IChemE form, that are generally used in specialist circumstances. The IChemE (www.icheme.org) describe their forms of contract as performance-based contracts for the design and construction of a process plant and other output-based projects. They specifically address the liabilities and testing regimes appropriate for a performance plant and are used extensively across a range of process industries. It is, therefore, rare for these types of forms to be applicable to ground investigation works.

5.4.   Contract parties

Figure 5.1 shows some typical relationships between various parties that have contractual responsibilities within a ground investigation contract.

5.5.   The purpose of contracts

A contract should facilitate the procurement of a suitable investigation to

- enable assessment of ground conditions
- enable acquisition of ground data for design and future asset management
- save on future build costs
- identify, determine, reduce and allocate risk
- obtain information that minimises the impact of the future development, including sustainability objectives.

A contract should clearly define responsibilities, including suitable provisions for responsibilities under CDM, such as the roles of Principal Designer and Principal Contractor. The contract should

Figure 5.1 Typical parties to a ground investigation contract

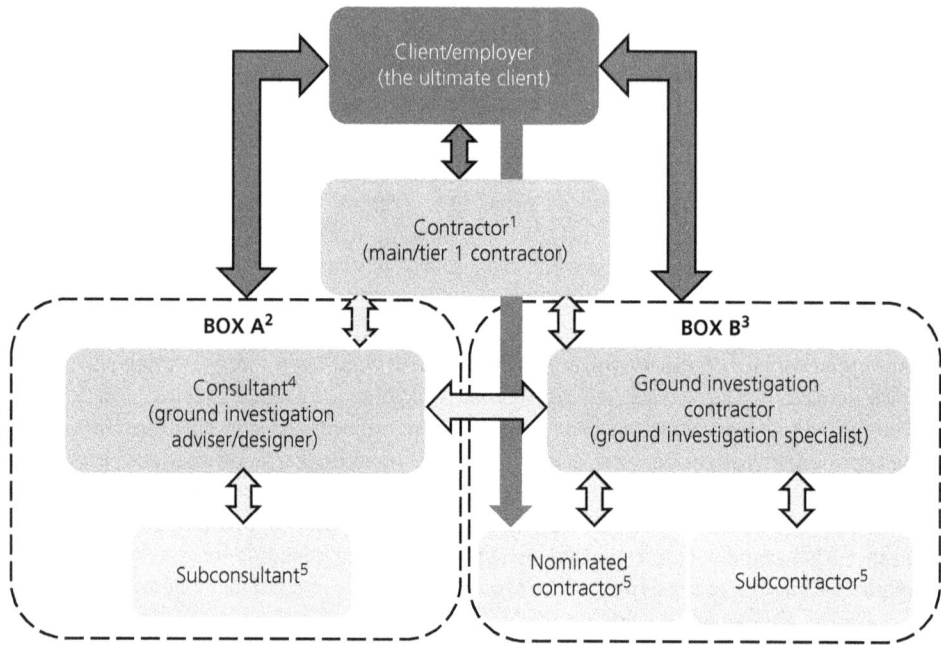

Notes:

1. A Contractor may be commissioned by the Client to project manage and sometimes procure the ground investigation. The Contractor may also be appointed as Principal Contractor. In these types of contract the contractual relationship between the Contractor and the consultant, who often acts as the Designer, can be unclear.
2. In Box A, the consultant may liaise independently with the Client/Employer but should also provide instructions to subconsultants and the Contractor on behalf of the Client. They should not provide instructions to subcontractors or other specialists employed by the Contractor.
3. In Box B, the ground investigation contractor is responsible for liaising with any subcontractors that they employ and any nominated subcontractors, if identified in the contract. The ground investigation contractor may take instructions from the Client, Contractor (if contracted) or consultant, but the contract should make it clear where responsibilities lie between those two parties.
4. The consultant will often act as the Client's representative and may be formally appointed as Principal Designer or Designer and, if contracted by the Client to employ the ground investigation contractor, will act as the employer under the ICC Conditions of Contract. Often the consultant will provide the Investigation Supervisor.
5. Subconsultants, subcontractors and nominated contractors will be a combination of specialists or organisations which can provide additional resources not available during the contract.

clearly define the responsibilities for the quality and technical objectives of the ground investigation, which could involve an 'Investigation Supervisor' or a party with similar delegated powers.

The Client/Employer should design elements of a contract at the early stages. Figure 5.2 sets out the general process for contract procurement and the most suitable conditions of contract for each stage of the process and the activities being procured.

Figure 5.2 Route to efficient contract procurement of ground investigations

Key Activity	Applicable Conditions of Contract	Client Legal Obligations	Notes
Project Inception		Client to understand roles and responsibilities. Create a competent team. Start to compile pre-construction information. Set up the Health and Safety File.	Commercial Clients must notify the HSE. Domestic Clients may appoint the Contractor or Principal Contractor or Principal Designer to notify the HSE.
Appoint Ground Investigation Adviser	NEC Professional Services RIBA Professional Services	Ensure the competence of the individual/ organisation. Ensure that the Health and Safety File is made available to all parties that are appointed.	May be an internal specialist or an external consultancy.
Appoint Designer or Principal Designer	NEC Professional Services RIBA Professional Services	Ensure their competence. Make appointment, preferably in writing. Define their powers and responsibilities. Complete pre-construction information	If there is more than one Designer a Principal Designer must be appointed. Designer is likely to be the best equipped to prepare the specification and act as Investigation Supervisor.
Appoint Principal Contractor	NEC ECC all options available Civils Engineering Contractor's Association Joint Construction Tribunal Institute of Chemical Engineers Federation Internationale des Ingénieurs Conseil Supply of Goods Client/Contractor own terms	Ensure their competence. Make appointment, preferably in writing. Clarify who is contracting with the Contractors and Designers. Develop Construction Phase Plan. Provide appropriate welfare facilities. Submit F10 to HSE, if notifiable.	Ground investigation is classed as construction and is part of the total construction activities. Unless a ground investigation is a standalone activity for feasibility purposes, a scheme will normally be notifiable.
Appoint Ground Investigation Contractor	NEC ECC Option B Infrastructure Conditions of Contract NEC ECC Short Contract	Ensure their competence and those of all other Contractors (subcontractors). Discuss specification requirements and compliance of the bid/ offer. Check level of supervision. Ensure site specific RAMS are provided. Check compliance of plant and machinery.	Direct contracting with the ground investigation contractor can provide many benefits. Ensure the ground investigation contractor has the required personnel and experience to act as Principal Contractor.
Project Completion		Ensure the Health and Safety File is complete.	Upon completion the geotechnical data must be made available in electronic format to other interested parties e.g., Designers, Facilities Management Contractors, Remediation Contractors, Piling Contractors etc

5.6. Ground investigation contract differences

The procurement of ground investigations is different from the procurement of works and services during the later stages of a scheme, one of the most fundamental differences being that accommodation of change is expected (even encouraged) and the identification of change to the ground model can be when the benefits of the investigation are most appreciated. As such, although the deliverables of a ground investigation may have been specified, including exploratory hole logs, tests, reports, monitoring data and so on, the content of the logs (e.g. the ground or groundwater conditions) cannot be anticipated. This can lead to additional scope, deeper boreholes, increased monitoring periods and so on.

Unlike the prearranged quantifiable deliverables of the construction stages, the findings of a ground investigation can lead to additional costs and changes to the programme that may not have been allowed for. To better understand a site, a PCI document (as explained in Chapter 3, Section 3.3) and desk studies are essential to provide an initial ground model (including risks associated with historic changes etc.) and understanding of the potential underlying ground conditions and antici-pated ground model. PCI serves as a basis for developing construction programmes and sched-ules. Understanding site-specific constraints and conditions enables the delivery team to plan and sequence activities more effectively, leading to improved project timelines. However, until a ground investigation is completed, unforeseen ground conditions should be anticipated. Therefore, ground investigation contracts need to be flexible to change and offer protection to the Client and Contractors alike.

Ground investigations are typically undertaken during the early stages of projects and involve examining and gathering information about the ground and groundwater conditions beneath a site. In contrast, design focuses on creating plans and specifications for the project. The information concerning the underlying ground and groundwater conditions is used to identify constraints pos-ing a risk or limitation to the project. The design incorporates mitigations to reduce the risks posed by the hazards to the project. In certain cases, ground investigations can be used to eliminate risks by abandoning further development of a project in a particular location in preference to other more suitable locations.

It should be noted that ground investigations do of course have their limitations. Even with the most thorough expansive scope it is not possible to achieve complete certainty of the underlying ground or groundwater conditions and there is the inherent risk of unforeseen ground conditions. The ground investigation contract should encourage the Contractor and Designer to work together to reduce the risk of unforeseen ground conditions and increase the identification of known unknowns, although this inherent risk further demonstrates why ground investigations can only ever be seen as providing data to advise Designers on the ground conditions.

5.7. Contract content
5.7.1 Introduction
As discussed in Chapter 5, Section 5.6, ground investigations are unsurprisingly about 'investigat-ing' the ground and Clients should reflect this in the conditions of contract that they adopt. While the contract forms a legally binding agreement between the Client and the Contractor, it should be developed to accommodate change and uncertainty. Therefore, conditions of contract should be chosen that have sufficient flexibility to accommodate the nuances of the ground investigation and to include contents which are appropriate to the successful management of the work.

5.7.2 Phasing of works
The contract should consider whether programming and sequencing of works are important in the effective and safe completion of the ground investigation. This is dependent on the scheme – for instance, a single phase of ground investigation may be suitable for a 'low-complexity', single discipline scheme. Conversely, ground investigations may be undertaken during many phases spanning years for large-scale, multidiscipline schemes which may require multiple works contracts.

Contracts should clearly identify the proposed phasing of a ground investigation, and these should be cognisant of the proposed works and unknowns raised during revisions of the ground model, design constraints, construction and any planning restrictions.

Potential phasing of works may include

- desk study including site walkover
- non-geotechnical surveys – for example, topographical, archaeological, ecological, arboricultural
- utility surveys – that is, PAS 128 (BSI, 2022).
- unexploded ordnance surveys (UXO)
- nonintrusive investigations – that is, surface geophysical surveys
- intrusive investigations (this may in itself take place as several phases)
- monitoring.

5.7.3   Involvement of other specialists

The identification of risks that may require specialist services should be identified in the PCI document provided by the Client to the Contractor, including any desk-based studies. These may include environmental requirements, archaeology or UXO. If this is not well understood at the time of contract development there are sources of information that can be reviewed to understand the potential risk to the scheme, including Environment Agency resources, MAGIC mapping (DEFRA, 2024) and CIRIA C681 *Unexploded Ordnance (UXO): A Guide for the Construction Industry* (Stone *et al.*, 2009).

If proceeding with intrusive works is considered too high a risk during initial phases of the scheme, preliminary ecological surveys and nonintrusive surveys using specialist techniques (magnetometer, ground-penetrating radar (GPR) etc.) should be considered in the contract to allow intrusive works to proceed, noting that these have the potential to impact the progress of intrusive works.

5.7.4   Specific items for inclusion in a ground investigation

The following matters need to be considered when preparing a ground investigation contract

- identification of the contractual roles and responsibilities of the Client (Employer, Principal Contractor etc.), the consultant (Engineer, Designer, Investigation Supervisor) and the party undertaking the ground investigation works
- clear definition of site access/traffic management/security provisions and responsibilities
- unavoidable damage/reinstatement obligations
- physical conditions or artificial obstructions that could not reasonably have been foreseen
- artesian groundwater
- testing (field (in situ) and laboratory)
- sample quality
- mineral deposits, mine shafts/adits, voids
- contamination
- objects of value/historical interest (i.e. archaeological artefacts, scheduled ancient monuments).

5.7.5   Site welfare and safety provisions

Accommodation (including welfare) and safety provisions should be part of the contract for the duration of the works as part of the legal and contractual responsibility of the Contractor.

The scale of accommodation and safety provisions in place for the works may be scaled up or down to match the activities being undertaken during the duration of the works or phases; this may include offices, canteen, lighting and other ancillaries. Throughout the life cycle of the scheme, welfare provisions should meet the requirement of the workforce, be fit for purpose and meet the minimum requirements set out in Schedule 2 of CDM 2015; for instance, are there sufficient toilet and washing facilities for the number of site staff, should showers be provided for particular activities, have welfare provision and cleaning activities been allocated?

Prior to developing the contract, the Client should be contacted to confirm whether they have specific requirements or recommendations based on their knowledge of the site (which could include office facilities, welfare, personal protective equipment (PPE), access requirements, provision of defibrillators etc.). The Client's knowledge and understanding of the site is crucial in the risk management of a contract, as detailed in Chapter 3.

Where a site is extensive in size – for example, for a long linear length scheme (highway/rail) – consideration should be given to satellite welfare facilities local to any remote location. Where an investigation is on a construction site or within an existing development, welfare facilities, cleaning activities and safety measures may be provided by the Principal Contractor or Client; in this case, it is important to make it clear where the responsibility for providing suitable facilities lies.

5.8. Special contract considerations
5.8.1 Rail contracts (access, rail corridor etc.)
Planning ground investigation works on railway projects can come with its own unique challenges that, if not considered during the planning stages of the ground investigation, can significantly impact final delivery and costs.

Figure 5.3 Working during a possession with rail-approved specialist plant. Courtesy of Chris Wade, Soil Engineering Geoservices Ltd (AGS photography competition entry)

Items to be considered with contracts for rail-related ground investigations should include

- appointment of rail-trained operatives and supervisory staff
- use of rail-approved plant
- location of the site operations (i.e. lineside, high street, depot)
- access including planning of possessions, approved routes
- working hours compliance
- local knowledge
- delays or late 'hand backs'.

5.8.2 Road contracts

Works on existing highways and those for the construction of new roads have differing requirements to be considered within a contract.

Works in the existing highway may be commissioned from two sources, namely

- the Highway Authority itself (the party responsible for the operation and maintenance of the road, the County Council, Combined Authority, National Highways, Transport Scotland etc.)
- a private party doing work on the highways (e.g. a developer building a new access roundabout). These works would normal be carried out under a Section 278 agreement of the Highways Act 1980 (HMG, 1980) with the relevant highway authority.

Works in the existing highway will require appropriate traffic management (TM). The contract should make clear who is responsible for the provision of this TM and any constraints stated (e.g. working hours, safety constraints and access restrictions during poor weather). The requirements to provide TM on high-speed roads are more onerous and should be undertaken by an appropriately certified traffic management provider. The contract should make clear under what circumstances a Contractor may be entitled to claim for any delays in working caused by TM and where it is the Contractor's responsibility.

Any road reinstatement required should be to the appropriate standard and may require sign off by the highway authority. The works will require personnel to hold the relevant New Roads and Street Works Act 1991 qualifications (NRSWA) (HMG, 1991).

The construction areas of new roads, particular those that are not part of a wider development, such as housing or an industrial estate, are generally longitudinal in nature and require a ground investigation along a narrow corridor of land. Typically, the ground investigation will be done in two stages.

- The first stage will prove the route option (selecting the best road alignment).
- The second stage will be much more intensive to support the detailed design of the works.

Usually, these works are likely to take place before the Client has ownership of the land. The longitudinal nature of the project means it is likely that multiple landowners will be affected. Access to the land for the purposes of ground investigation requires separate negotiation and agreement. This is normally best undertaken by the Client (or their representatives) who will hold an existing relationship with the landowner. These agreements will take time to implement and should be allowed for in any programme of works. When drawing up the contract it is important to clearly define who is responsible for these agreements. It is also imperative that the Contractor is informed of any details of the access agreements and undertakes the works in full compliance with those arrangements. Protection of surfaces and property (buildings, fences, crops etc.) is paramount. These investigation works will take place prior to the construction of the project and the performance of the works (good or bad) will impact the Client's relationship with the landowners through the main construction works.

5.8.3   Work over and near water

When procuring ground investigations that involve work over or near water, it is imperative to thoroughly evaluate the experience and capability before finalising the selection of the Contractor.

Clients should request bidders to submit a comprehensive logistical plan for accessing the over-water operation as part of their tender returns. This should include details about the working platform, how crew access will be gained, its safe operating conditions, details of the welfare and first aid provision on board, specifications for any proposed vessels, their capabilities, and relevant certifications. The roles and responsibilities of key personnel should be clearly defined, including the barge master or marine superintendent provided by the Contractor, especially regarding safety-related matters and who will make the final decisions.

Preconstruction information, as detailed in Chapter 3, Section 3.3, should be provided in advance to potential bidders to enable them to select suitable, safe operating equipment. Some of these items are included in the Site Information section of NEC contracts.

Weather contingencies should be considered to minimise disputes arising from weather delays that could impact the project schedule; clear criteria should be established during the tender stage for safe operating conditions regarding wind speed, visibility and wave height. It should be clearly stated that any delays due to these conditions will be considered a Client risk, given that the Client specifies the project timeline and selects the site.

Work over or near water often necessitates additional licenses and permissions compared to land-based investigations. The procurement process should mandate the timely acquisition of these licenses. Typically, this responsibility should rest with the Client or should be facilitated on the Client's behalf before the Contractor's appointment to minimise delays post award. Work may require an application to the Environment Agency for an environmental permit for flood risk activities. This will usually require collaboration between the Client and Contractor.

5.8.4   Work over Coal Authority assets and other mineral workings

Clients, Designers and Contractors must comply with the Borehole Sites and Operations Regulations 1995 (HMG, 1995). This includes notifying the HM Inspector of Mines of all boreholes exceeding 30 m in depth and located within 1 km of past or present mineral extraction activities. Additionally, a Coal Authority permit is required when drilling through geological units containing coal that has been or might be exploited. The requirements for each of these and who is responsible for compliance should be clearly stated in the specification and contract.

If exploratory holes are likely to intersect, disturb or enter any Coal Authority property interests (e.g. unworked coal, coal workings, shafts and adits), an application must be submitted to the Coal Authority for permission to conduct the investigation.

Figure 5.4 Dynamic sampling work near water. Courtesy of Tom Androsiuk, EPS Ltd (AGS photography competition entry)

The application should include a site plan showing the boundary, location, number and depth of any planned boreholes or excavations. It should also detail any proposed grout mixes for treatment works, full drawings and calculations if treating mine entries, and any previous works such as coal mining risk assessments, desk studies or borehole logs. Documentation and guidance are available on the Coal Authority website.

Clients must agree to the Coal Authority's non-negotiable permit terms and conditions. These include providing an indemnity to the Coal Authority against any liability for claims, losses or damages arising from any works carried out on the site by the Client, its agent, representative, Designers or Contractors. This indemnity is valid for a period of 12 years and should be completed by the Client, landowner or a party that will have control over the land for that period. The terms and conditions transfer liability for any harm whatsoever arising out of the operations onto the applicant. The Client may attempt to secure terms with its Contractors/consultants that are 'back-to-back' with the applicant's obligations to the Coal Authority under the Terms and Conditions. The AGS guide *Drilling into Coal Authority Assets: A Review of The Coal Authority's Terms and Conditions for Entering or Disturbing Coal Authority* (AGS, 2009) recommends that the terms should be avoided by consultants and Contractors because many of the obligations impose strict liabilities which will not be covered by professional indemnity insurance or require services that are not within the consultant's or Contractor's areas of expertise.

It should be noted that obtaining a Coal Authority permit typically takes four weeks. Therefore, this time should be factored into the planning of any site start dates.

5.9. Working with the specification
5.9.1 Contract compatibility
It is important to use contract terms that work with the specification. The *UK Specification* includes a standard bill of quantities and measurement rules that can be used with most contract conditions

used for ground investigations. Unless the contract specifically excludes the remeasurable aspect of this specification, it will be incompatible with lump sum and professional service type contracts.

5.9.2   Programme requirements

Various formats may be demanded for programmes, with Microsoft Project, Primavera and simplified Excel versions being common. The contract should explicitly specify the required format, along with the submission and acceptance timeframes for updates.

Ideally, programmes should incorporate resource loading, predecessor and successor tasks/activities and, on request, be presented on a hole-by-hole basis aligned with Schedule 2 of the *UK Specification* which should provide all exploratory hole and in situ test information. If there are known access constraints or specified sequences, this information should be supplied to facilitate programme development. In the absence of such restrictions, a logical sequence for hole formation is assumed.

The accuracy and completeness of Schedule 2 are critical in developing the fieldwork element of the programme. For instance, if the depth of individual holes is not detailed with respect to constituent parts (cable percussion and rotary) or if specific in situ testing and installation requirements for each hole are not outlined, accurate duration estimates become challenging. Therefore, attention to Schedule 2 details is vital for programme precision.

5.9.3   Method of measurement

The ground investigation contract should seamlessly link the method of measurement with the contractual requirements and the specification. Remeasurement of cost items is an essential part of a ground investigation contract that allows for flexibility and efficient acquisition of data. 'Lump sum' contracts or those where there is a fixed scope of works are rarely favourable for an effective investigation. Remeasurement can frequently lead to a reduced cost for the Client, where exploratory holes may be deemed unnecessary or are terminated early during the works.

It is important that the contract clauses do not conflict with the method of measurement. NEC Option A contracts are not remeasured, while Professional Service Contracts have a schedule of costs and, therefore, are not compatible with the ICE Specification for Ground Investigation unless significant changes are made.

5.10.   Risk management in contracts

One of the main duties of the Client is to ensure sufficient time and resources are allocated to a contract, particularly for health, safety and impact on the environment. There should be a fair apportionment of risk between contracting parties clearly stated within the contractual terms. Some of the main risks to be considered in contract documents are as follows

- health and safety considerations
- environmental risks
- insurance terms
- site security
- preconstruction information (PCI) informing parties of all known risks
- early warnings (in NEC contracts)
- pre-award and commencement meetings, including ECI for larger contracts
- risk identification and reduction

- clauses covering the consequences of negligence by the Contractor, Client and other controlling parties
- risk management processes, including supervision, risk register (in NEC)
- provision of service information (ideally 90 days) or other processes for protection of services
- delay costs/liquidated damages
- provision of performance bonds, retention etc.
- sustainability objectives
- weather: time of year, exceptionally adverse, 1 in 10 years, near shore environment, visibility, wind speed etc.
- rail costs
- traffic management costs
- limit of liability
- equitable enforcement of indemnities
- liability for 'damage' (including exceptions for unavoidable damage)
- access agreements, land rental, planning constraints, local objections
- ambiguities and inconsistencies
- hierarchy/priority of contract documents, reliance on documents
- force majeure events
- Z clauses (in NEC) and changes to common forms, deletion of Client/Employer responsibilities (including indemnities as above).

Various forms of contract apportion different levels of risk between Clients and Contractors. These are considered in Figures 5.5 and 5.6.

The NEC4 six Main Options A to F enable Clients to select a procurement strategy and payment mechanism most appropriate to the project. An illustration of the general apportionment of risk of the NEC options between Clients and Contractors is presented in Figure 5.5. There may be some movement of risk, depending on whether the Client takes the risk of inaccuracies in the bill of quantities in Options B and D and the extent of risk allocation in target contract options depending on how the 'share profile' is defined.

Similarly other forms of contract provide various levels of risk apportionment. Figure 5.6 shows some of the most common forms used for the procurement of ground investigations, which are summarised in Chapter 5, Section 5.2.

Figure 5.5 Financial risk of NEC main options (after Hughes, 2019)

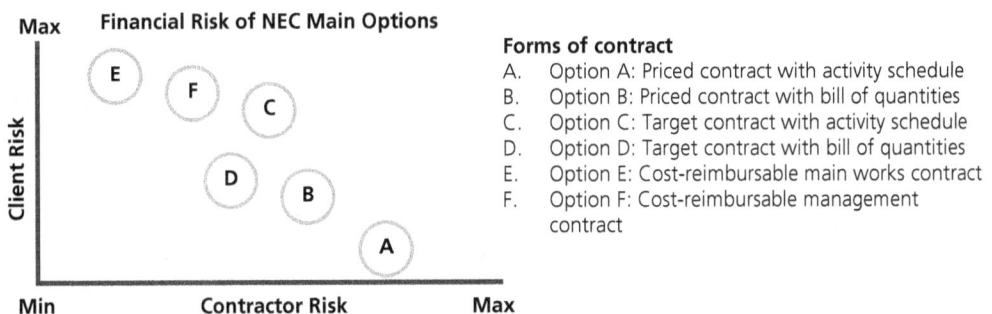

Forms of contract
A. Option A: Priced contract with activity schedule
B. Option B: Priced contract with bill of quantities
C. Option C: Target contract with activity schedule
D. Option D: Target contract with bill of quantities
E. Option E: Cost-reimbursable main works contract
F. Option F: Cost-reimbursable management contract

Figure 5.6 Financial risk for other contract options

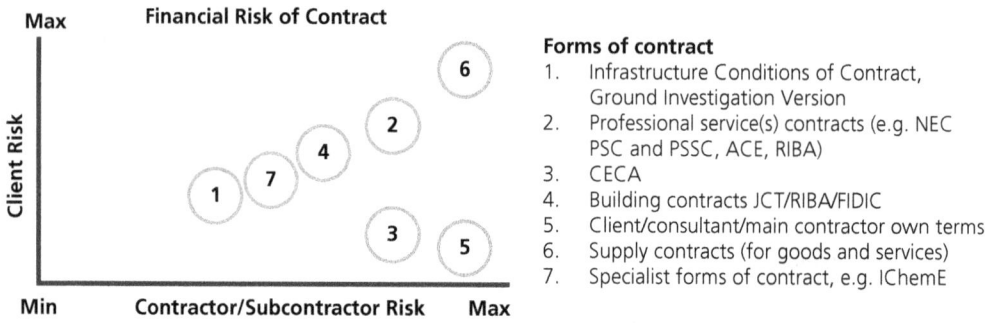

Forms of contract
1. Infrastructure Conditions of Contract, Ground Investigation Version
2. Professional service(s) contracts (e.g. NEC PSC and PSSC, ACE, RIBA)
3. CECA
4. Building contracts JCT/RIBA/FIDIC
5. Client/consultant/main contractor own terms
6. Supply contracts (for goods and services)
7. Specialist forms of contract, e.g. IChemE

5.11. Insurance requirements

Insurance requirements are an important part of a Contractor's obligations and these often determine the levels of liability cover for the Client. The most common forms of insurance that are applicable to a ground investigation are Employer's liability, public (and products) liability and professional indemnity (to a lesser extent as there is often no Contractor design or advice).

Other forms of cover often required are as follows

- machinery/plant: for Contractor's own and hired in plant
- vehicle fleet
- specialist – airside, over water, nuclear, lifting operations, pollution (including specialist asbestos cover in certain circumstances)
- Contractor's 'all risk'/contract works insurance (more relevant to construction activities).

Each of these insurances should have the required cover stated within the contract. This will include cover limits – for example, whether the insurance is for each and every claim or aggregate, inclusion of indemnity to principal clauses.

5.12. Conflict avoidance

Conflict can be defined as

> *a serious disagreement or argument, typically a protracted one.*
> Oxford English Dictionary, Oxford Languages 2024

Where a collaborative approach is taken and the contract encourages 'mutual trust and cooperation', the chance for conflict or disagreement is invariably minimised. Conflict between parties should always be avoided as it can cause damage to business relationships and brand reputations. They can also be costly and slow to resolve.

The Client's aim should ultimately be for their projects to not have any conflict arising between the delivery team. The reality is not necessarily so straightforward; conflicts are typically a result of disagreements which can be many and varied but often centre on fair payment – for example, agreement of final measure of the works, underpayment or late payment.

During or after a contract or specified works, there may be several times or opportunities where conflict or disputes could occur. Being proactive and transparent will assist in conflict avoidance but overall keeping dialogue and discussion open always helps this process. There should be a clear process in place to carry out quick and thorough assessments of any conflict risks. Contracts such as the ICC and NEC have mechanisms in place to recognise and deal with many types of conflict. Where conflict may occur that is not covered by the contract process or where contracts are chosen that do not recognise conflict, the Client should provide an agreement at commencement of the project which provides a procedure to resolve any conflict and a route to resolving any conflict fairly.

Appendix F shows various situations which may cause conflict during a ground investigation contract. The list is not exhaustive but gives some potential scenarios.

REFERENCES

AGS (Association of Geotechnical and Geoenvironmental Specialists) (2009) *Drilling into Coal Authority Assets: A Review of The Coal Authority's Terms and Conditions for Entering or Disturbing Coal Authority*. https://www.ags.org.uk/item/lpa-40-drilling-into-coal-authority-assets/ (accessed 09/10/2024).

AGS (2022) *UK Specification for Ground Investigation*, 3rd edn. Emerald/ICE Publishing, Leeds, UK.

BSI (2022) PAS 128:2022: Underground utility detection, verification and location – Specification. BSI, London, UK.

DEFRA (Department for the Environment, Food and Rural Affairs) (2024) MAGIC Map (Multi-agency geographic information for the countryside. https://magic.defra.gov.uk (accessed 09/10/2024).

HMG (Her Majesty's Government) (1980) Highways Act 1980. The Stationery Office, London, UK.

HMG (1991) New Roads and Street Works Act 1991. The Stationery Office, London, UK.

HMG (1995) Borehole Sites and Operations Regulations 1995. The Stationery Office, London, UK.

HSE (Health and Safety Executive) (2015) Construction (Design and Management) Regulations. The Stationery Office, London, UK.

Hughes K (2019) *Understanding the NEC4 ECC Contract. A Practical Handbook*. Routledge, Abingdon, UK.

Kennedy K (2024) NEC option X29 – a positive step to tackling climate change, but not without its risks. *AGS Magazine*, 20 May.

NEC (New Engineering Contract) (2017) *NEC Contracts*. NEC, London, UK. https://www.nec-contract.com/ (accessed 09/10/2024).

Stone K, Murray A, Cooke S, Foran J and Gooderham L (2009) *Unexploded Ordnance (UXO): A Guide for the Construction Industry*. CIRIA, London, UK, CIRIA C681.

FURTHER READING

NEC (2022) *Understanding the Use and Benefit of Performance Bonds in NEC Contracts*. NEC, London, UK.

NEC (2023) *Comparison of the Engineering and Construction Short Contract (ECSC) with the Engineering and Construction Contract (ECC)*. NEC, London, UK.

NEC (2023) *Guidance on using NEC Contracts for Ground Investigations*. NEC, London, UK.

RICS (Royal Institution of Chartered Surveyors) (2013) *Construction Security and Performance Documents*. RICS Professional Guidance, RICS, London, UK, guidance note GN 101/2013.

Association of Geotechnical and Geoenvironmental Specialists
ISBN 978-1-83662-029-7
https://doi.org/10.1108/978-1-83662-026-620251006

Chapter 6
Effective specification

6.1.   Introduction

Chapter 2, Section 2.2 introduced the three key elements to be considered when procuring ground investigations, namely cost, time and quality, and highlighted that effective procurement is finding the balance between these elements. All of these have to be considered when developing the scope of the work. The scope should provide a clear set of requirements (objectives) for the work and these should be formally outlined to the Contractor. An essential part of this process is the development of a fit for purpose specification, which can be defined as

> an act of identifying something precisely or of stating a precise requirement

or

> a detailed description of the design and materials used to make something.

In relation to engineering and ground investigation the specification is

> a detailed description of the dimensions, construction, workmanship, materials etc. of work done or to be done, prepared by an architect, engineer but more usually the Designer.

A specification is the document that describes in words what cannot be visualised or explained on a drawing or model. This is not only applicable to the construction industry but also to the ground investigation industry. In ground investigation, the specification can cover everything, including

- site establishment
- contract type
- timescales and delivery criteria
- systems and product quality
- applicable standards and how they are executed
- specific techniques to be used.

The type of specification should relate to the project and can be simple in form or very complex depending on project requirements.

Specifications

- are required for ground investigations
- are required during the design stage
- form part of the contract documentation
- play a key role in project fulfilment.

6.2.   Importance of a specification

There are many reasons why specifications are crucial to the ground investigation element of the construction process.

- The specification provides clear instructions on project intent, performance and ground investigation requirements.
- It references the quality and standards that should be applied.
- Any field and laboratory investigation techniques are clearly defined.
- Any installation and monitoring requirements are identified.
- It eliminates the need for information overload on the drawing or model, making identifying information easier.
- A specification can support project costing, not only the materials and products but also the performance and workmanship.
- Along with the drawings, the specification forms part of the contractual documents, helping minimise project risk and providing support should there be any legal disputes.
- It supports Client brief interpretation and gives the Client assurance that their commissioned investigation is the one being delivered.
- It is essential for the ground investigation phase of the project and links into the construction phase. It helps to provide clarity and records the thought process that produced the design information for management design of structures.
- By being clear, concise and information-rich, a specification provides answers to many onsite investigation questions, saving the project team, Client and Contractor time and money and reducing ambiguity or misinterpretation.
- After project completion, various parties can incorporate best practices and lessons learnt into a future specification, improving efficiency, providing quality assurance and ensuring project consistency.
- Specifications developed by individual companies and organisations also save the design team time and money because they can be developed over time and adapted to suit a project's specifics, drawing on specialist knowledge when needed.
- The specification is a living document to be used by the entire project team throughout the investigation phase; its value does not end at the ground investigation phase.
- Along with any variations or value engineering, it becomes a part of the project audit trail and a crucial part of the handover documents, forming the basis for asset management, asset maintenance, and even feeding into staff training and human resources policies.

6.3.   Developing the specification

Specifications used to be one of the last items written before issuing a tender package; however, this has changed over time, especially since the advent of various software packages including BIM (building information modelling).

Current best practice is to begin specification development as early as possible in the project life cycle. Early-stage specifications can capture information from the Client review, documenting the project objectives and making sure that this translates into an appropriate and effective ground investigation. Early development of the ground investigation specification also provides a better understanding of the delivery requirements for the ground investigation contractor. A thorough understanding of the Client requirements by the Designer, used to inform the scope of the specification, together with early contractor involvement (ECI), will help to ensure that the ground investigation output data provides all the information needed for subsequent design.

In essence, specification is about data communication and information exchange between the Client, Designer and ground investigation contractor. As project data requirements become more complex and we strive for more collaboration, it is crucial everyone has the same basis for project requirements – making specifications as essential now as they have ever been.

It is essential that the specification is fully informative and explains how the works are to be conducted to fulfil the objectives required by the investigation. If procurement is going to be effective, the specifier should be well informed and competent.

The person or persons drawing up the specification need to fully understand the required objectives of the ground investigation to produce a technically competent specification.

The specifier should

- have suitable experience, an understanding of the methods to be used and knowledge of procedures and best practice
- have a knowledge of construction methods to be employed and the geotechnical construction process
- understand the parameters required for design and how to obtain them
- understand ground conditions and geology
- have a sound knowledge of investigation methods and their limitations in the geology present at the site
- have knowledge of environmental requirements and obligations
- understand contractual requirements
- understand the obligations for health, safety, welfare, wellbeing and the environment
- understand timescales – which should be realistic both for the tender process and the delivery of the investigation outputs.

The level of skill required by the specifier will depend on the complexity of the points above – for example, to provide a specification for a two-storey residential development on a relatively uniform site with groundwater some distance below the depth which will be stressed by the structures could be carried out by someone with significantly less experience than for the specification of a complex infrastructure project with complex geology and groundwater within the zone which might be stressed.

Where environmental issues are paramount someone on the SiLC (Specialist in Land Condition) register might be considered as an adviser and specifier of the investigation and testing requirements, although it is likely that they would work in partnership with a geotechnical engineer.

The person drawing up the specification may well be someone with just a few years' experience. However, it is essential that if this is the case, the documents are assessed by a fully experienced individual prior to the document being issued.

Some investigations require complex methods and skillsets across several disciplines; this may require specifiers from different disciplines to collaborate in the writing of the specification.

The RoGEP (Register of Ground Engineering Professionals) scheme provides a means of assessing whether an individual has appropriate skill and knowledge to compile the specification.

This can provide Clients with the assurance that works are being guided by someone with demonstrable ability. Generally, someone with the attributes of specialist or adviser level would be considered to have the appropriate skills to write the specification, although this may be limited by their specific experiences which may not be sufficiently broad to be able to fulfil the whole process; thus others with the appropriate experience and skills may need to support or contribute to the process.

6.4. Specification contents

If procurement is going to be effective, the specification should be unambiguous, well developed and clear about the aims of the investigation.

The specification requires to be compiled such that it provides prospective tenderers with sufficient information in order to price the proposed works while understanding the limitations and constraints of the site and methods to be employed. To do this the specifier should as a minimum have at their disposal a desk study.

If a desk study has not been compiled, then this should be the first step in the procurement process. Following this, the specification should be updated using the findings of the desk study.

Ideally the desk study should form part of the specification and, therefore, should be provided to the tenderer along with the specification. The importance of the desk study cannot be overstated: it should detail the past history of the site and provide details of geology, topography, water courses, access constraints, utilities information, potential contamination risks, ecological and environmental hazards and constraints, and any other elements which might need to be considered when working on the site (e.g. UXO risk, archaeological remains, ecological constraints).

With this information, along with a knowledge of the proposed works and an understanding of the required outcomes, the specification can be drawn up. The specification also requires knowledge of the methods associated with the expected geology and understanding of a realistic timescale for the investigation deliverables.

The specification should ideally contain the following

- description of the site location and topography
- description of the proposed development
- description of the expected geology (this may include historical/previous surveys)
- site location plan
- description of expected investigation methods
- exploratory hole locations plan
- a schedule of exploratory holes (referred to as Schedule 2)
- details of sampling required giving type and spacing
- details of any field (in situ) testing required
- details of any installations, monitoring and sampling regimes
- levels of supervision to be provided
- laboratory testing expected (both geotechnical and environmental)
- a bill of quantities based on the *UK Specification* (AGS, 2022)
- details of the expected programme for works.

The document should ideally follow the structure of the *UK Specification*, noting any divergence from the specification.

6.5.   The importance of adequate supervision

The *UK Specification* (AGS, 2022) recognises the importance of adequate supervision and provides items within the bill of quantities where supervisory staff beyond those who are legally or contractually obliged to be provided can be specified. These individuals may include supervisors for particular specialist activities and Level 3 qualified advanced drillers. Importantly the specification can be used to define the role of the Investigation Supervisor.

BS 5930 (BSI, 2020) states that the Investigation Supervisor

> should have suitable qualifications, skills and specialist experience in ground investigations. They should also be familiar with the purposes of the particular investigation and be suitably skilled and experienced in the specific methods of investigation required.

A list of the typical duties of the Investigation Supervisor is provided within the *UK Specification* but, above everything, the Investigation Supervisor must provide '*independent technical direction and oversight of the ground investigation*'. The Investigation Supervisor can be from an independent organisation or come from either a Contractor or Designer organisation and should be formally appointed by the Client. This appointment, preferably in writing, should also include a set of clearly defined delegated powers.

The main benefits of selecting an appropriate Investigation Supervisor include

- Client reassurance and clear communication channels
- oversight for managing geotechnical risk for the project (i.e. CD622) and wider CDM aspects
- ensure that the ground investigation is undertaken in accordance with the specification and complied with explicitly
- review of detailed design decisions/input if undertaking validation investigation during construction phase
- ability to react to changing ground conditions, which may not have been foreseen when scoping the investigation, such that in situ testing/material sampling can be immediately implemented*
- maintenance of programme by proactive management avoiding unnecessary standing time charges – for example, avoidance of delays in decision making regarding instrumentation installations or early termination of any exploratory location due to unforeseen circumstances
- Client support in respect of approval of Contractor's invoices/applications for payment or work measures
- providing valuable site knowledge which is beneficial when compiling the testing schedules
- providing valuable site knowledge which is beneficial when compiling the ground investigation report
- providing valuable site knowledge for geotechnical feedback reporting, lessons learnt and wider learning.

*This requires that the Client gives the Investigation Supervisor the authority to instruct variations to the original scope of works, which is likely to have a commercial implication.

6.6.   UK Specification for ground investigation

As stated throughout this publication, the adoption of the *UK Specification* (AGS, 2022) is recommended where possible as it addresses the aspects raised in Chapter 6, Section 6.1. The *UK Specification* has been developed to improve the effectiveness of the scope of ground investigation works and bring it up to date with developments in techniques.

Some of the main revisions from the previous second edition include

- site surveys – addition of pre- and post-condition site inspections
- advanced technologies – addition of items for UXO surveys, use of UAVs and drones, cavity inspections and using remote methods
- underground utility surveys – recognising PAS 128 (BSI, 2022) (NB the term 'services' replaced with 'utilities')
- professional attendance – removal of this term and replaced with specified roles
- competence – inclusion of registers – for example, RoGEP, SiLC
- drilling methods – additional items for sonic drilling, measurement while drilling (BS EN ISO 22476-15: BSI, 2016) and core scanning
- casing – included as a sperate item
- laboratory testing – abortive laboratory preparation of test specimen added and range of tests increased to include advanced testing methods
- reporting – revised with inclusion of a factual element of a ground investigation report (GIR); the term 'interpretative' has been removed
- contamination testing – to BS 10175 (BSI, 2017) and Land Contamination Risk Management (Environment Agency, 2020)
- innovation – recognises advances in ground investigation techniques and equipment but also provides flexibility to introduce innovation.

6.7.   Method of measurement
6.7.1   Introduction

The method of measurement is how the ground investigation work will be measured and costed as the investigation proceeds. It should be carefully thought out to ensure clarity of cost around different issues that may arise during the investigation. A clear and concise method of measurement will enable an accurate cost to be developed allowing the Client to understand whole-life costings for any ground investigation, irrespective of size and complexity. It will ensure that all parties understand the ongoing and final costs and will therefore help avoid conflict. The use of a standard method of measurement will also provide a level playing field for all tenderers so that a like-for-like comparison of tender returns can be undertaken.

6.7.2   Bill of quantities

As part of the method of measurement, bills of quantities are produced to provide the ground investigation contractor with a specific schedule of items which require prices to be allocated. They typically comprise a set of preliminary items for establishment of costs, reporting and supervision, together with scheduled items for the exploratory work (comprising drilling, sampling, in situ testing, installations) as well as post-site laboratory testing. The bill of quantities is a key component of any tender and its accuracy will be of paramount importance in avoiding any contractual disagreements, conflicts or compensation events. The bill of quantities should fully incorporate the deliverables within the specification and schedules and align with the goals of the intended ground investigation.

6.7.3 Methods of measurement

The recommended method of measurement is to adopt the bill of quantities set out in the *UK Specification*. These bills of quantities have been specifically developed to address the uncertainties that arise due to the nature of ground investigation works and to reduce risk to all parties.

Historically there are several other examples of methods of measurements which are used. Two of these occasionally used documents are

- Manual of Contract Documents for Highway Works: Volume 4: Bills of Quantities for Highway Works (Standards for Highways, 2009)
- CESMM4: Civil Engineering Standard Method of Measurement (ICE, 2019).

Neither of these methods of measurements are suitable for detailed ground investigations and often expose both the Client and Contractor to additional risk when key costs are omitted due to the specification and bill of quantities being mutually exclusive.

6.7.4 Format of the bill of quantities

The format of the bill of quantities should be in accordance with the one presented within the *UK Specification* for ground investigation. This will ensure that tenderers are working from a standardised format and that responses are easily scrutinised to select the most advantageous bid.

The preamble to the bill of quantities should also be provided and allow a concise list of item coverage for the bill of quantities. This avoids any additional caveats or assumptions which may complicate the tender evaluation process.

Amendments to the standard layout should be avoided.

The bill of quantities should be prepared in entirety by the team compiling the specification and other contract documentation. Each section of the bill of quantities should reflect the full scope of the ground investigation, and the tenderer should be encouraged to add any additional items required to complete the scope of services if unintentionally missed.

The bill of quantities should reflect the specification, schedules and the drawings. It should be accurate and error free.

6.7.5 Worked example

A Client wishes to procure a ground investigation contractor to undertake a geotechnical investigation for a new structure. The following is an example brief which may accompany the tender document.

> Undertake the drilling of six boreholes to a nominal depth of 10m or rockhead whichever is shallower, collect samples, complete in situ testing, off-site laboratory analysis and prepare a ground investigation report (GIR) on the factual data obtained.

Below is an example of a poor bill of quantities based on the information above using a separate method of measurement.

Item	Description	Quantity	Units	Rates	Total
	BILL 100 – PRELIMINARIES				
1	Mobilisation of equipment and subcontractor prelims	1.00	SUM		
2	Site investigation works to inform the design – boreholes	6.00	NR		
3	Sampling and testing	1.00	SUM		
4	Laboratory testing/reporting	1.00	SUM		
				TOTAL TENDER	

The above example has only four line items to enable the Contractor to price in accordance with the scope. In this example the precise scope of works and methodology cannot easily be interpreted and will potentially lead to a wide range of Contractor's assumptions.

Using the *UK Specification* bill of quantities, the scope and methodology can easily be populated on a greater range of line items. It follows a more structured approach and enables accurate remeasurability, with any final account based on actual work completed. For this example project, the bill of quantities would be split into seven sections as detailed below.

Section A: General items, provisional services and additional items

This will cover the arrangements for storage and welfare, mobilisation of drilling plant (cable percussion in this example), supervision, as-built drawings, site specific method statements and risk assessments, reporting (GIR) including digital AGS data and reinstatement.

(continued on next page)

Number	Item description	Unit	Quantity	Rate	Amount £
A	**General items, provisional services and additional items**				
A1	Establish on site offices, stores and amenities for the Contractor	sum	1		0.00
A2	Maintenance of offices, stores and amenities for the Contractor	week	1.5		0.00
	Establish on site all machines, equipment and utilities for a GREEN category site				
A4	Establish on site and remove from site, one cable percussion boring machine and ancillary plant and equipment	nr	1		0.00

82

Continued

Number	Item description	Unit	Quantity	Rate	Amount £
A20	**Provide project management, supervision and execution personnel**				
A20.3	Site Agent (specify grade required)	p. day	7		0.00
A20.10	Site Logger (specify grade required)	p. day	7		0.00
A21	Establish the location and elevation of the ground at each exploratory hole	nr	6		0.00
A22	Preparation of health and safety documentation including risk assessments	sum	1		0.00
	Reporting				
A35	Digital copy of the draft ground investigation report (or specified part thereof)	sum	1		0.00
A37	Digital copy of a final ground investigation report (or specified part thereof)	sum	1		0.00
A43	Interim issue of digital data in AGS data format	nr	1		0.00
	Temporary works and other matters				
A66	Reinstatement using cold bituminous material surfacing	m3	3		0.00
	Total Bill A carried to summary				**0.00**

Section B: Cable percussion boring

This will detail the method of advancement of the boreholes, number and depth of boreholes, and any backfilling requirements. This will also cover advancing through hard stratum and standing time of plant and equipment.

(continued on next page)

Number	Item description	Unit	Quantity	Rate	Amount £
B	**Cable percussion boring**				
B1	Move cable percussion boring machine, equipment and crew to the site of each borehole and set up	nr	6		0.00
B4	Advance borehole between existing ground level and 10 m depth	m	60		0.00
B5	As item B4 but between 10 and 20 m depth	m	Rate only		
B10	Advance borehole through hard material or obstruction	hr	3		0.00

Continued

Number	Item description	Unit	Quantity	Rate	Amount £
B13	Standing time for cable percussion boring machine, equipment and crew	hr	6		
	Backfill cable percussion boreholes				0.00
B18	Backfill borehole with bentonite pellets	Lin.m	60		0.00
	Total Bill B carried to summary				**0.00**

Section D: Pitting and trenching

This will cover any hand-dug inspection pits for service avoidance and breaking out of surface obstructions.

Number	Item description	Unit	Quantity	Rate	Amount £
D	**Pitting and trenching**				
	Inspection pits				
	cable percussion boring/dynamic sampling				
D1	Excavate inspection pit by hand to 1.2 m depth	nr	6		0.00
D2	Extra over item D1 for breaking out surface obstructions	hrs	3		0.00
	Total Bill D carried to summary				**0.00**

Section E: Sampling during formation of exploratory hole

Section E will detail the type and quantity of samples required. Sampling quantities can be calculated based on the *UK Specification* for sample frequency or to suit the testing regime for the proposed scheme.

Number	Item description	Unit	Quantity	Rate	Amount £
E	**Sampling during formation of exploratory hole**				
E1	Small disturbed sample	nr	60		0.00
E2	Bulk disturbed sample	nr	18		0.00
E6	Open-tube UT100 sample – by cable percussion or dynamic sampling	nr	24		0.00
	Total Bill E carried to summary				**0.00**

Section H: Field testing

This section will detail the in situ testing to be undertaken. In this example this will be limited to standard penetration testing.

Number	Item description	Unit	Quantity	Rate	Amount £
H	**Field testing**				
H1	Standard penetration test in borehole between ground level and 10 m depth	nr	24		0.00
H2	As item H1 but between 10 and 20 m depth	nr	Rate only		0.00
	Total Bill H carried to summary				**0.00**

Section K: Geotechnical laboratory testing

Section K will detail the requirement for quantities of geotechnical testing.

Number	Item description	Unit	Quantity	Rate	Amount £
K	**Geotechnical laboratory testing**				
	Classification				
K1	Water content	nr	24		0.00
K2	Liquid limit (four-point method), plastic limit and plasticity index	nr	24		
K11	Particle size distribution by sieving	nr	18		0.00
K12	Sedimentation by pipette	nr	18		
	Chemical and electrochemical				0.00
K17	Sulfate content of water extracted from soil	nr	6		0.00
K25	pH value	nr	6		0.00
	Shear strength (total stress)				0.00
K89	Unconsolidated undrained triaxial compression without the measurement of pore pressure (set of three specimens)	nr			
K90	As item K89 but for a single specimen	nr	24		0.00
	Total Bill K carried to summary				**0.00**

Summary of bill of quantities

A concise summary page gives the summary costs from each section detailed above.

Summary of Bill of Quantities

		£
A	General items, provisional services and additional items	0.00
B	Cable percussion boring and dynamic sampling	0.00
C	Rotary drilling	Not used
D	Pitting and trenching	0.00
E	Sampling during intrusive investigation	0.00
F	Dynamic probing and cone penetration testing	Not used
G	Geophysical surveys	Not used
H	Field testing	0.00
I	Instrumentation	Not used
J	Installation monitoring and sampling	Not used
K	Geotechnical laboratory testing	0.00
L	Geoenvironmental laboratory testing	Not used
M	Technical support from ground practitioners or other personnel	Not used
N	Long-term storage of samples	Not used
	Tender total:	0.00

6.7.6   Importance of the preamble

The preamble is an important introduction to the bill of quantities for the ground investigation tender document. In the *UK Specification*, the preamble is also referred to as Annex 1 'Bill of Quantities for Ground investigation'. The preamble identifies the subheadings and item descriptions for the work covered by the respective items and is split into 25 subheadings which define the item coverage for each section of the bill of quantities.

The preamble enables the opportunity to clearly define what should be covered in line items within the corresponding bill of quantities. This could include additional allowances for specialist plant for enabling access, obtaining permits and licences or other specialist tests or services not generally covered in a standard ground investigation contract – for example, pontoons for overwater work.

6.7.7   Recommendations on content

Recommendations to be considered when developing the bill of quantities are provided in Table 6.1. These recommendations are intended as a guide to producing a bill of quantities which provides control on costs as the investigation proceeds.

It is recommended that ground investigation should be priced on a remeasurable basis due to the unknown nature of the ground conditions (whether an additional phase of work) and the flexibility required by all parties to amend scope, testing requirements and durations to suit actual site conditions. The bill of quantities in the *UK Specification* has been specifically laid out to facilitate this.

Table 6.1 Recommendations for item usage (continued on next page)

Item	Recommendations
Preamble	Use the preamble to clarify the bill of quantities as far as possible so that aspects are not missed.
A items	Include clarification on what is expected in items A1 and A2 of the bill of quantities, these being the items that cover offices and stores for the Contractor. Items A4 to A12 cover establishment on site of all plant, equipment and services. In previous editions there has been a lump sum attributed to each of these, despite the complexity of planning and costing that goes into them. It is recommended that Contractors make their breakdown of these item costs as clear as possible.
Plant, equipment and utilities	Note that the *UK Specification* indicates that the 'Designer' shall provide an initial estimation of the plant, equipment and utilities necessary to complete the works and provide itemisation in Bill A. This has not traditionally been the 'norm' and perhaps should be highlighted as outlined in the NEC guidance note that the Contractor should/shall check these and, where necessary, change within the bill to suit. Alternatively, the Designer could leave these blank and state that these are for the Contractor to determine, although this may lead to inconsistencies between multiple tender returns.
Lump sums	The use of lump sum items should be avoided and, instead, time-related terminology used, such as per hour, per day, per week and so on. This is of particular importance when defining supervision requirements whereby ground investigations are supervised by geotechnical specialists, site agents and drilling supervisors, and these are charged under a day rate.
Day rates	Include a table of typical day rates/set-up types.
Programme	Think about the programme for the ground investigation. Discuss a realistic programme with Contractors during pre-tender visits to make sure duration of the ground investigation is understood as far as possible between all parties. Allow the Contractor to adjust numbers of days of professional attendance in A7 to suit their programme.
Supervision and technical staff	Supervision and technical input should be reasonable and proportional to size and complexity of the investigation.
Minimum requirements	Clearly specify the minimum requirements.
Contract specific items	Use the bill of quantities to identify any special or unusual requirements or provisions that are unique to that project by including contract-specific items at end of each section of the bill of quantities – for example, trackway, temporary works, specialist items (e.g. LIDAR, INSAR, specialist monitoring equipment etc.) and nominated laboratories/suppliers.
Units	Ensure unit of measurement is applicable to each item and discuss with the specialists.
Provisional sums	Minimise use of provisional sums (especially irrelevant ones).

Table 6.1 Continued

Item	Recommendations
Rate only	Include sufficient use of rate-only items if there is a possibility they may be needed. If items are to be required, avoid rate-only and estimate or calculate the quantity required.
Standing time	Define what is included in standing time allowance to avoid conflict. Identify whether additional works outside of the scope are to be charged under this item or identify whether 'dayworks' or additional rates are required.
Contamination	Include relevant items and associated quantities should contamination be expected.
Health and safety	Items for consideration to comply with CDM 2015.
Technical support	Correct use of Bill M and Bill N; include estimated quantities or state not needed.
Phases	Any special sequencing of activities/testing should be clearly specified in the schedules and relevant items added to the bill so that time/resources/equipment/pricing can be done accurately.

6.7.8   Associated recommendations to improve the measurement process

Site visits are recommended with the Designer, Contractor and Client at the start of the tender process to facilitate understanding of the bill of quantities requirements. This enables the Client to meet with the Contractors to discuss and clarify the project objectives, proposed techniques as well as possible alternative approaches. The discussions should be open and frank and include: access, equipment, techniques, alternative proposals, programme, communication and data management.

Where suitable, there should be a facility in the bill of quantities to add an alternative or value engineering proposal – for example, an alternative or better technique or time-saving initiative. This is particularly important to allow innovation in the supply chain to not only potentially offer cost savings but also provide a greater degree of certainty to the data produced.

Contractors should provide a careful list of assumptions made to help benchmark the costs and progress in the fieldwork.

A prestart meeting on site is recommended to ensure that all parties, including the Contractor's site personnel, are aligned on the purpose of the ground investigation, technical aspects, programme, communication, costs and reporting.

Good communication methods need to be established on site to ensure that standing time is understood and agreed quickly at the time. Similarly, good communication will ensure that changes in quantities are understood and agreed. A collaborative approach will achieve the best outcome for all parties.

A means for regular updates on progress and cost is recommended, monitored against the bill of quantities, with any differences from initial assumptions highlighted.

REFERENCES

AGS (Association of Geotechnical and Geoenvironmental Specialists) (2022) *UK Specification for Ground Investigation*, 3rd edn. Emerald/ICE Publishing, Leeds, UK.

BSI (2016) BS EN ISO 22476-15: Geotechnical investigation and testing. Field testing - measuring while drilling. BSI, London, UK.

BSI (2017) BS 10175:2011+A2:2017: Investigation of potentially contaminated sites. Code of practice. BSI, London, UK.

BSI (2020) BS 5930:2015+A1:2020: Code of practice for ground investigations. BSI, London, UK.

BSI (2022) PAS 128: Underground utility detection, verification and location. Specification. BSI, London, UK.

Environment Agency (2020) Land Contamination Risk Management (LCRM). UK Gov, Open Government Licence v3(0).

ICE (Institution of Civil Engineers) (2019) *CESMM4: Civil Engineering Standard Method of Measurement*, 4th edn. Emerald/ICE Publishing, Leeds, UK.

Standards for Highways (2009) *Manual of Contract Documents for Highway Works: Volume 4: Bills of Quantities for Highway Works*. https://www.standardsforhighways.co.uk/tses/attachments/1e56dda3-cc10-4588-90be-a1e9a8a7870a?inline=true (accessed 09/10/2024).

FURTHER READING

Highways England (2020) *Design Manual for Roads and Bridges. Managing Geotechnical Risk.* Highways England, London, UK, CD 622.

Association of Geotechnical and Geoenvironmental Specialists
ISBN 978-1-83662-029-7
https://doi.org/10.1108/978-1-83662-026-620251007

Appendix A
Principles of prevention for ground investigations

Appendix A (continued on next page)

No.	Principle of prevention	Description	Examples
1	**Avoid risks**	The most effective way of avoiding risk is to entirely eliminate the hazard that gives rise to risk in the first place.	▪ The best way to eliminate a risk associated with a piece of work is not to do the work. Therefore, in a ground investigation, only do the work that is required to achieve the technical objectives: don't break ground unnecessarily, don't investigate to a greater depth than is needed. ▪ Arrange for shutdowns such as isolations in high voltage electricity environments or possessions during rail works. ▪ Plan investigation locations to be away from known hazards – for example, utilities, other infrastructure.
2	**Evaluate risks that cannot be avoided**	Where identified risks cannot be avoided, assess the likelihood and potential severity of their impacting directly on the works to be undertaken.	▪ For cable percussion boring, the weight of tools and, hence, the manual handling risk is, in part, dependent on the borehole diameter. Consider what minimum borehole diameter will be necessary to complete the works to enable the manual handling risk to be assessed.
3	**Combat sources of risks**	Risks should be addressed, or dealt with, at the source. In other words, the control measures should be as close as possible to the hazard and effective in reducing the risk.	▪ Control dusts at source with water or vacuum suppression, such as wetting down haul routes or wet cutting as against dry cutting. ▪ Noise control at source through effective plant maintenance reduces the level of noise emitted compared with noise curtains and hearing protection.

4	Adapt the work to the individual	In some ground conditions, frequently specified sampling and testing regimes can result in significant, repetitive manual handling for operatives that can be a source of fatigue. This is particularly an issue in wet, muddy environments that are common in the ground investigation industry. Just like physical limitations, people have limits when it comes to their ability to analyse cognitive data, such as instructions or instrumentation readings. Monotonous work can be alleviated by adapting the design of workplaces, the choice of work equipment and the choice of working and production methods.	This may not be relevant to most GI procurement situations, but the specifier should be aware of steps that can be taken, such as reducing the number of samples bagged together to reduce the weight to be carried or reviewing the height at which items are stored within storage containers to ensure that everything is within easy reach of the site staff.
5	Adapt to technical progress	Technology continues to influence the modern workplace, so it is important to keep informed about the latest technical knowledge when selecting working methods, equipment, materials and work equipment. In general, technical progress leads to improved performance, better ergonomics and reduced risks.	■ Technology has led to video and wireless systems that can remotely inspect confined spaces such as sewers or remotely monitor instruments, eliminating the need for personnel to enter potentially hazardous environments. ■ Specify mechanical methods to lift and carry heavy loads and reduce manual handling e.g. tele-handlers, forklifts, rod handlers, hi-ab lorries. ■ Ensure that the programme allows a balance of resources – for example, drilling rigs standing due to limited number of sets of specialist in situ test equipment such as geophysics and packer testing.
6	Replace the dangerous with non or less dangerous	Commonly known as 'substitution', this principle involves reviewing the choices that are available and then selecting the ones that either pose no danger to workers or selecting the ones where the dangers are reduced. A broad sweep of hazards in the working environment, the task, materials, plant and tools should be considered. Substitution can also involve changing the proposed process.	■ Using low-temperature asphalt can help reduce risk of burns when carrying out road repairs. ■ CPTs are quicker than conventional drilling so involve less work time, which may be appropriate in particularly hazardous environments. ■ Combine techniques – for example, geophysics in preference to intrusive investigation techniques – to reduce the number of exploratory holes or manual handling. ■ Replace oil-based substances with water-based substances.

Appendix A (continued)

No.	Principle of prevention	Description	Examples
7	**Develop a coherent prevention policy**	This could cover technology, organisation of work, working conditions, social relationships and more. This reinforces the fact that a company should take control of the management of itself and its Contractors and not leave it to an external body or consultant.	▪ Ensure the contractual environment allows for safe systems of work to be developed. ▪ Communication – inductions, briefings, lessons learnt. ▪ Carry out independent audits and inspections. ▪ Promote and maintain positive reporting practices (near miss, hazard spots etc.).
8	**Give collective protective measures priority**	Collective protective measures control the risks to more than one person and have major advantages over individual protective measures. Personal protective measures rarely prevent accidents from occurring. Instead, they may mitigate the outcome, but only if they are correctly fitted and worn. Personal protective measures only provide mitigations to those who are using them rather than the whole workforce.	▪ See also Principle 3 examples above. ▪ Ensure machines and equipment are independently inspected. ▪ Ensure machines have compliant guards. ▪ Ensure access to adequate onsite welfare. ▪ Ensure fall from height collective prevention (handrails and barriers) are used in favour of lanyards, inertia reels and so on.
9	**Provide appropriate instructions to employees**	Instructions should be communicated in a way that is readily understood by the workers. In other words, be both comprehensible and relevant. Common communication techniques include induction training, toolbox talks, daily briefings and providing written or pictorial instructions.	▪ Give daily briefings highlighting current and specific risks, attended by all parties and retain records. ▪ Regularly engage with the workforce to provide opportunities for feedback to help improve the process. ▪ Give clear and concise task specific instructions and briefings are provided to ensure safe systems of work are understood and followed. ▪ Be alert to the possibility of non-native English speakers among the work force.

Association of Geotechnical and Geoenvironmental Specialists
ISBN 978-1-83662-029-7
https://doi.org/10.1108/978-1-83662-026-620251008

Appendix B
Risk allocation – good practice and guidance

Appendix B (continued)

Item	Task/Issue	Risk Allocation				Comment
		Client	Contractor	Designer/ Investigation Supervisor		
Site identification	For example, site, adjacencies, ownership boundaries, constraints and access restrictions. Outline details of proposed development, where relevant.	A	–	I		Client brief and project direction needed to ensure requirements are incorporated.
Third party impacts	For example, access, disturbance, utility, rail, transport, maritime and party wall issues should be considered.	A	–	R		This work is done in advance of designing the ground investigation and carried out with the assistance of a Client's advisor (consultant/ Designer/Investigation Supervisor).
Desk study	For example, geology, contamination, asbestos, topography, groundwater, historical maps, flooding situation to inform the choice of proposed layouts and initial costings.	A	–	R		This work is done in advance of designing the ground investigation and carried out with the assistance of a Client's advisor (consultant/ Designer/Investigation Supervisor).
Additional searches related to site (other than ground conditions)	For example, UXO, services searches, archaeology, ecology and reports undertaken to inform the choice of proposed layouts and cost plan.	A	–	R		This work is done in advance of designing the ground investigation and carried out with the assistance of a ground investigation adviser (consultant/Designer/Investigation Supervisor).
Site visit	Typically, essential to a proper understanding of the site and potential constraints.	A	C	R		This work is done in advance of designing the ground investigation and carried out with the assistance of a ground investigation adviser (consultant/Designer/Investigation Supervisor).

Design of ground investigation and contract

Item	Description	A	R	Guidance
Defining the investigation objectives	For example, investigation scope appropriate to the site and the project being proposed on it.	A	R	The Designer incorporates project requirements into the relevant contract documentation following the Client's brief and code/legislative requirements.
Notification of or permission from relevant authorities	For example, relevant permissions may be required from relevant statutory bodies – for example, Highways England, National Rail, Coal Authority, Local Authority, Environment Agency (groundwater and landfills, waste)	A	R	The Designer incorporates project requirements into the relevant contract documentation. In some cases, obtaining permissions may be a task assigned to the contractor, particularly where they are more familiar with the process than the Designer.
Access to the ground investigation locations	For example, identified routes for trafficking by site investigation vehicles may be necessary to limit damage to the ground or to observe agreements with third parties.	A	R	The Designer incorporates project requirements into the relevant contract documentation.
Risks inherent in land being investigated	For example, land use, interference with other activities, livestock, water and proximity to water courses, flooding, proximity to over/underground services or apparatus.	A	R	The Designer incorporates project requirements into the relevant contract documentation.
Environmental issues that may be impacted by the proposed works	Ecology (protected species, invasive plants), archaeology, contaminated land.	A	R	The Designer incorporates project requirements into the relevant contract documentation.
Controlled environments	Special measures may be required to access areas where diseases such as foot and mouth, swine vesicular, bird flu are identified.	A	R	The Designer incorporates project requirements into the relevant contract documentation.

Appendix B (continued)

Item	Task/Issue	Risk Allocation			Comment
		Client	Contractor	Designer/ Investigation Supervisor	
Infectious diseases (potential for)	Spread of infection, social distancing, safe working, vulnerable receptors.	A		R	The Designer incorporates project requirements into the relevant contract documentation together with government or local authority guidance. This includes for diseases such as foot and mouth, tuberculosis and so on in wildlife and livestock, as well as humans.
Existing conditions and reinstatement requirements	For example, can vary enormously from residential gardens and lawns to reinforced concrete or highway. Clarification on who is responsible for reinstatement and to what standard is required.	A		R	The Designer incorporates project requirements into the relevant contract documentation.
Need to install instrumentation and monitor	For example include consideration of the need to access periodically and protection of the installation against damage/vandalism	A		R	The Designer incorporates project requirements into the relevant contract documentation. Client/landowner/agent to support.
Position, type and depth of investigation	For example, consideration should be given to any enabling works likely to be required to access proposed GI locations and the practicalities of doing the work. Early contractor involvement may be valuable in this case.	A		R	Specialist activities may require a degree of early contractor involvement to arrive at an achievable outcome.

Adequacy of ground investigation	The design of the investigation will need to ensure that the outcome provides sufficient information to satisfy project and code design and reporting requirements	A	R	The Designer incorporates project requirements into the relevant contract documentation.
Unexpected ground conditions	Conditions may differ from expectations, requiring change to scope of investigation or programme or aims not met as intended.	A	R	The Designer incorporates project requirements into the relevant contract documentation.
CDM project set-up	CDM aspects including appointment of Principal Designer and Principal Contractor (where applicable), project notifications as required. Responsibility for preparation of preconstruction stage information pack for Contractors	A, R	C	The designer makes the Client aware of their responsibilities and incorporates project specific arrangements into the relevant contract documentation.
Special contract requirements	For example, Client permit to work requirements, archaeology dust monitoring needed, ecological survey requirements (i.e. bats, newts, protected species, birds nesting) and need for a watching brief by a qualified professional.	A	R	The Designer incorporates project requirements into the relevant contract documentation.
Contract additional conditions	For example, need for collateral warranties, ownership of reports, confidentiality of information or other requirements	A	R	The Designer incorporates project requirements into the relevant contract documentation.
Specification	Standard *UK Specification* is generally recommended, together with relevant British Standards (or as applicable locally).	A	R	Specialist specifications or requirements may require early contractor involvement to arrive at an achievable outcome.

Appendix B (continued)

Item	Task/Issue	Risk Allocation			Comment
		Client Contractor	Designer/ Investigation Supervisor		

Letting the contract

Item	Task/Issue	Client Contractor	Designer/ Investigation Supervisor	Comment
Provision of preconstruction information (PCI)	In accordance with Regulation 2 of the CDM Regulations.	A, R —	—	Where some or all of the required PCI is missing, the Client is required to 'take reasonable steps to obtain the information needed to fill any gaps identified by, for example, commissioning relevant surveys'.
Appropriately qualified and experienced personnel proposed by the Contractor	To ensure that project or general requirements are met and for quality assurance.	A R	—	Clarified and confirmed prior to the letting of a contract to the preferred Contractor.
Adequate resources proposed by the Contractor	To ensure that project or general requirements are met and for quality assurance.	A R	I (R)	Clarified and confirmed prior to the letting of a contract to the preferred Contractor.
Appropriately accredited laboratories and testing arrangements, equipment	To ensure that project or general requirements are met and for quality assurance.	— A, R	(R)	Clarified and confirmed prior to the letting of a contract to the preferred Contractor.

Adequate programme to complete the work	To ensure that project or general requirements are met and for quality assurance.	A	R	(R)	Clarified and confirmed prior to the letting of a contract to the preferred Contractor.
Negotiation and agreement to proceed with the works	Pre-contract meeting usually held with preferred bidder to clarify issues.	A	(R)	R	Usually clarified and confirmed prior to the letting of a contract to the preferred Contractor.

Contract stage and fieldworks

Construction phase plan - CDM	Requirement of CDM Regulations to allow proper planning of the works.	A	R	C	Required before any work on site is initiated
Management and planning of all site activities in relation to health, safety and wellbeing, safety of third parties	Following health and safety legislation, regulations and targeting best practice.	A	R	C	Arrangements need to be in place prior to starting on site. RAMS documents reviewed and agreed before starting the relevant activity on site. Welfare facilities must be accepted by the Client prior to start on site.
Technical details – to be provided - including method statements or temporary works designs as applicable.	To develop the methodology and confirm details ahead of the work.	A	R	C	In accordance with the contract. Where the contract documents require specific details on methodology proposed, as may be the case with propriety systems or more complex testing methods, this allows the Contractor to confirm how the specification requirements will be achieved.
Preconstruction survey	To review the practicalities of carrying out the investigation at the site and, where necessary, record precondition of exploratory hole locations.	A	R	(R)	To be undertaken in advance of mobilising to a particular site.

Appendix B (continued)

Item	Task/Issue	Risk Allocation			Comment
		Client	Contractor	Designer/ Investigation Supervisor	
Security of site (site compound and investigation locations/ equipment), signage	Protester action, vandalism, obstruction and so on.	A	R	C	In accordance with the contract.
Notice to stakeholders	As required.	A	R	(R)	In accordance with the contract.
Traffic management and liaison	As required.	A	R	C	In accordance with the contract.
Access to the site and GI locations; notification to relevant bodies, obtaining consents	While licenses/permits may have been agreed prior to the contractor appointment, there is often still a requirement to obtain specific consent to access a site on a particular day/time.	A	R	C	In accordance with the contract.
Weather conditions and working conditions	Inclement weather causes issues with progress, access or health, safety and welfare.	A	R	C	In accordance with the contract.
Unexpected issues	The nature of investigation works is that they can encounter the unexpected.	A	R	C	For example, archaeology, ecology (birds nesting in rig).
Adequacy of plant equipment and resources	Consequential hazards arising from inappropriate methods used .		A, R	(R)	Creating or destabilising geohazards – for example, unstable ground, sinkholes, landslips, pollution of aquifers, gas emissions from underground workings, striking utilities, settlement of trial pits or backfilled boreholes.

Risk	Description				Guidance
Unexpected impacts from unrelated works on site	For example, impact on access to work areas.	A, R	R	C	Client to resolve.
Compliance with all recent standards	Ground investigation works are covered by a range of standards and industry guidance, which must be adhered to.		A, R	(R)	Contractor's works to cover .
Quality assurance, quality control and nonconformances	Evidence is required to confirm that all works have been completed to the appropriate standards or to clearly indicate where there have been deviations from the standards.		A, R	(R)	Contractor's works to cover.
Technical supervision of the workforce	Appropriately qualified and experienced technical supervision is critical to maintaining the quality of a ground investigation.		A, R	C	Contractor's works to cover.
Technical oversight	Ensuring that an investigation is meeting its aims and that the scope of works responds to the ground conditions being discovered by the investigation as it progresses requires competent technical supervision.		A, (R)	R	Client appointee – for example, Investigation Supervisor role.
Keeping records and daily reporting	This is normally a requirement of the specification for the works being undertaken		A, R	C	Contractor's works to cover. Timely provision of records enables effective management of the investigation.
Programme maintenance and changes	As required.	A	R	(R)	Providing, updating, keeping to.
Liaison with stakeholders and other relevant third parties	As required.	A	R	(R)	Contractors to lead following initial agreements by Client.

Appendix B (continued)

Item	Task/Issue	Risk Allocation			Comment
		Client	Contractor	Designer/Investigation Supervisor	
Technical queries, early warnings and notifications	As required.		A, R	(R)	Contractors in accordance with the contract/specification.
Reinstatement to specified requirements	Requirements will depend on the ground surface at the site and the land use e.g. highway reinstatement, livestock areas etc.		A, R	C	Contractor's works to cover in accordance with the specification. Reinstatement requirements may be established by regulation as well as landowner requirements.
Adequacy of post-condition surveys	Must be sufficient to identify changes as a result of the investigation works.		A, R	C	Contractor's works to cover good pre- and post-condition surveys, providing an important record of any damage caused by the investigation works and thus enabling liability for any claimed damage to be assigned.
Sign-off of reinstatement by relevant parties	Agreement from the landowner/affected parties that land has been returned in an acceptable condition.		A, R	C	As per specified requirements. It is advisable to undertake a post-works inspection with all parties and stakeholders to agree satisfactory completion of the works and return of the site to landowner/occupier.

Laboratory testing and reporting of factual information

Item	Task/Issue	Risk Allocation			Comment
		Client	Contractor	Designer/Investigation Supervisor	
Scheduling of required lab testing	Generally undertaken by the Designer/Investigation Supervisor but may be jointly with contractor.		(R)	A, R	Generally, the Designer/Investigation Supervisor will schedule the appropriate testing to support the design. However, if the Contractor is also the Designer (D&B contract) then this may not be the case.

Item	Description				Guidance
Undertaking and resourcing	Managing the process to meet agreed programme requirements.		A, R	(R)	Contractor's works to cover.
Quality assurance and quality control	Compliance with all standards and agreed limits of detection and so on.		A, R	(R)	In accordance with accreditations or as specified.
Sample handling/chain of custody	Track sample location and avoid deviations in samples – for example, due to time from sampling to testing.		A, R	(R)	Contractor's works to cover.
Contamination/asbestos protocols	Health, safety and wellbeing within lab or transport of samples.		A, R	(R)	Contractor's works to cover.
Preparation of GI factual report and corresponding digital AGS data	For example, within project programme timescales.		A, R	(R)	Contractor's works to cover.
Technical comments on draft report and testing	Usually carried out by the Investigation Supervisor or other technical advisor.		A, (R)	A, R	The Contractor will be expected to incorporate these comments into the final report.
Acceptance of GI report	Commercially, completion of an investigation requires a formal acceptance of the factual report.	A	(R)	R	By Client or Investigation Supervisor.
Disposal of samples from investigation	To appropriately licensed facility.	A	R	C	As specified.
Retain report copy and data for identified timescale	There may be industry or legal requirements setting a minimum retention period.	A	R	(R)	

Finalisation of report and contract close out

Item	Description				Guidance
Storage of ground investigation report and digital data (AGS)	For example, version control and maintenance of record copy.	A	(R)	R	Generally, Client responsibility but usually contract will specify period for Contractor to retain records.

Appendix B (continued)

Item	Task/Issue	Risk Allocation			Comment
		Client	Contractor	Designer/ Investigation Supervisor	
Lodge GI data with BGS or other body as appropriate as required by the Client.	Best practice is that all ground investigation data should be deposited (in AGS format) with the BGS, contributing to the publicly available geological record.	A		R	As per specified requirements. Encouraged to further the industry's understanding of ground-related risks.
Prepare geotechnical baseline report (GBR)	GBRs may be required but are typically not undertaken as part of the ground investigation.	A		R	Generally initiated by the Client unless there are other contractual provisions.
Interpretation and establishment of parameters for engineering design; development of site geotechnical/ eological models	The objective of a ground investigation is to produce data that can then be used to support the design of the proposed development.			A, R	By appointed Designer/Contractor if interpretation included in contract.
Share GI information as may be relevant e.g. piling contractor (AGS) and compile within CDM health and safety file for the site	The data from the ground investigation should be shared with all parties that require the data; relevant data must be included in the CDM health and safety file by legislation.	R		A	Generally initiated by the Client unless there are other contractual provisions ('Principal Designer' has primary responsible for health and safety file under the legislation). It is important that the digital data (AGS and other) is shared fully with all relevant parties.

Association of Geotechnical and Geoenvironmental Specialists
ISBN 978-1-83662-029-7
https://doi.org/10.1108/978-1-83662-026-620251009
Emerald Publishing Limited: All rights reserved

Appendix C
Contract clauses and provisions

Appendix C (continued on next page)

Contract clauses and provisions	Definitions and statements of use	Comments
Limitation of aggregate liability	A limitation of liability clause in a contract is a clause that limits a party's liability to a commercially reasonable amount, such as the contract sum.	Contracts that do not contain a limitation of liability clause are likely to expose the party providing the services or goods under the contract to unlimited liability. Often the liability limits in a ground investigation contract are set at the extent of the insurance coverage of the Contractor. There is generally no measure of 'reasonable' limitation where the terms of a contract have been agreed between two parties of equal bargaining power, as noted in the Association of Geotechnical and Geoenvironmental Specialists Loss Prevention Alert 77. *Limits of Liability – Are They 'Reasonable' Under the Unfair Contract Terms Act?* (AGS, 2023).
		A contract without a limitation of liability clause exposes either party to significant risk. Take for example, a contract for party A (employer) and party B (contractor) to complete a ground investigation at an airport. If a services strike (gas) were to occur which resulted in the destruction of infrastructure and planes and hundreds of millions of pounds' worth of loss, party A will seek to recover that loss from B. However, if there is a cap on B's liability, say for £1 000 000, A could only recover that amount from B (subject to carve-outs or liability which cannot be excluded at law). The amount of the limitation on liability should be reasonable considering factors such as the contract price, the nature of the goods or services provided and the parties' other representations or commitments. It is important for both parties that the limit of liability does not exceed the insurance coverage, as a larger claim could result in bankruptcy.
		Generally, at law, liability for death or personal injury caused by negligence or fraud and fraudulent misrepresentation cannot be excluded/limited.
		Once agreed, a liability cap will not apply to the carve-outs. Using the above example, if agreed to, an aggregate cap of £1 000 000 will not apply if party B have been negligent, and death has resulted.

All businesses must limit the risks that they face. Agreeing to an aggregate cap on liability provides both the business and insurers some certainty of risk and clarity for the employer. Different parties may argue that a large business should carry the burden. However, to do so may mean charging an increased premium for services and products to adequately balance those risks. It is not, therefore, unreasonable to request such a limitation. It is incumbent on contracting parties to ensure that the liability cap is not disproportionate to the contract value wherever possible.

Limitation of liability wording that excludes certain provisions – for example, where there is an indemnity provision and the clause relating to the cap excludes that provision in the total cap – can sometimes have a limiting effect on the application and effectiveness of the cap on liability – for example, if there is an indemnity for breach of contract and a cap on liability for three times the contract value but excluding the indemnity clause, then there would be no cap for breach of contract.

For example, think of a simple purchase order for goods or services. The seller is obliged to provide services/goods of a certain quality and the buyer is obliged to pay for them. If the seller breaches his or her obligations to supply the services/goods of a certain standard and quality and the buyer has paid for services/goods that now require defect rectification, if the buyer pays a third party for defect rectification, it might seek to recover the costs of that defect's rectification from the seller as damages.

This principle does not require a specific contractual provision entitling a party to damages for there to be a liability for damages. This means that if a contract is silent on the issue of damages for breach, general law of damages will apply. A party must have suffered actual loss to be able to recover damages in this way and must be able to provide evidence substantiating the loss. Unless there is a specific contractual provision limiting liability, there is no limit to the amount of damages that may be recovered under general damages.

General damages

A general principle of common law is that if a party breaches a contract (i.e. fails to comply with any obligation it has under the contract) and as a result of that breach the other party to the contract suffers a loss, the party suffering the loss may (subject to certain legal principles beyond the scope of this guide) recover the amount of loss from the party in breach as damages. Damages for breach of contract are intended to put the claimant back in the same position as if the requirements of the contract had been performed.

Appendix C (continued on next page)

Contract clauses and provisions	Definitions and statements of use	Comments
Liquidated damages	Often referred to in construction contracts as delay damages or colloquially as 'LDs', these are damages that have a fixed sum predetermined by the parties. The principle of liquidated damages is that they are a genuine pre-estimate of the loss that will be suffered by a party to a contract in the event of a breach.	Liquidated damages are generally expressed as some sort of formula. The most common examples are
		▪ in construction contracts, a daily or weekly rate of damages if the event of failure to complete the works by the contracted date
		▪ in service contracts, a reduction of the monthly fee by a set percentage if contracted KPIs (key performance indicators) are not met.
		A liquidated damages regime only applies where there is a specific contractual provision to that effect. If there is no such provision, general damages will apply. Like general damages, liquidated damages require that a party breach its contractual obligations before liability arises. Unlike general damages, liquidated damages do not require that a party suffer loss, or be able to substantiate the loss, for there to be a liability. All that is required is that the specified breach occurred, and liquidated damages will automatically apply. Clear and realistic delivery schedules in the contract will minimise the likelihood of having to pay LDs.
		Liquidated damages are not always found in contracts, but are an acceptable form of regulation where they are commercially reasonable and it is stipulated that they are the sole and exclusive remedy for breach of certain obligations (e.g. failure to reach practical completion on time).
Combined damages	Claiming both general and liquidated damages need not be mutually exclusive.	There is some inconsistency in court decisions as to whether a contractual entitlement to liquidated damages will mean that the entitlement to general damages is no longer available. In most circumstances, a detailed contract that clearly sets out each party's rights and obligations will mean that liquidated damages will be the sole and exclusive right to claim damages where a breach giving rise to a liability for liquidated damages occurs.

Where there is a contractual indemnity that obliges one party to indemnify the other for any losses incurred arising out of a breach of contract, this may not be the case, as the contractual right to indemnity will operate separately to the right to liquidated damages and will override the general principles regarding damages.

It is, therefore, important in dealing with a contractual indemnity for breach of contract, where there is also a separate liability for liquidated damages, to ensure that the indemnity excludes any losses arising from breach of contract for which liquidated damages apply. A suitable clause for inclusion in the indemnity provisions is as follows.

'This indemnity shall not apply to breach of contract for which the remedy of liquidated damages is available under another provision of this contract.'

Leaving the schedule blank or inserting 'N/A', '£0' or 'Nil' does not mean that the principal is not entitled to unliquidated damages in another form.

In some jurisdictions, the courts have interpreted leaving the schedule blank or inserting 'N/A', '£0' or 'Nil' to mean that while liquidated damages do not apply, general damages (uncapped unless there is a separate limitation of liability) will still apply. Clear drafting is the key. If it is the parties' intention that no damages are payable for a particular breach, liquidated or general, then a clause to this effect should be expressly included in the contract. Similarly, if liquidated damages are specified and payable, they should be the sole and exclusive remedy for a particular breach and this should be clearly stipulated in the contract. A legal opinion is recommended where the contract could potentially lead to significant damages.

It is appropriate and reasonable for contracting entities to limit the risk in their contracts to ensure appropriate risk-reward and reduce the general contract costs by limiting the risks to the Contractor, including a 'per week' rate as opposed to 'per day'. The risk to the Contractor can also be limited both by the rate of accrual and by the inclusion of a total cap (typically at 10% of the contract sum).

Use of damages terms: 'NA, £0 or Nil':

Many contracts show the liquidated damages rate in a schedule to the contract. You might see liquidated damages expressed as 'N/A', '£0' or 'Nil' to indicate that liquidated damages are not applicable.

Appendix C (continued on next page)

Contract clauses and provisions	Definitions and statements of use	Comments
Consequential damages	Consequential, indirect and/or special damages refer to damages that do not flow directly from the breach/event.	Incurring such damages can have a disastrous financial impact, totally out of proportion to the nature of the contract. It is virtually impossible to include a contingency for such damages in the contract price because the exposure can be far in excess of the contract value.

The financial risk is difficult to predict and calculate and, therefore, cannot be reasonably insured against. Examples of consequential damages include loss of profit or revenue that an employer may claim as a result of delay to a project, loss of use of the premises due to late delivery or performance, loss of goodwill and reputation due to late delivery or performance, or the expense incurred by a tenant in leasing alternative space because the premises could not be used when planned. Imagine, for example, a football stadium that does not open on time – the loss of profits on the number of seats that would have been sold but for the delay alone would be substantial.

Unless specifically disclaimed or waived, liability for consequential damages exists in all contracts. The only way to avoid consequential damages is by using specific language that clearly disclaims liability for them. As stated above, such damages can have serious financial implications and therefore lead to higher tender prices. The most common type is lost profits and lost business opportunity. In the UK, these types of damages are recoverable unless there is a written clause in the contract that expressly excludes them. In other jurisdictions (like in Europe), the parties must expressly state which damages are being excluded in order for the clause to provide protection.

It is difficult to assess the financial risk of consequential losses and often agreeing to a 'mutual' exclusion in a contract is more acceptable to Contractors than a single exclusion in favour of either party. A separate 'carve-out' within any total aggregate cap on liability, by capping the consequential damages at a lower level but within the overall cap, could also limit the contractual risks. If the Contractor is unable to agree specific exclusion or carve-outs for these losses, but it has secured a total aggregate cap on its liability, then the Contractor will have some protection as this will be rolled into that total cap. |

Payment terms

Payment terms are the terms on which the Contractor agrees to be paid. This typically references the length of time the employer has from the date of the Contractor's invoice to make a payment.

Sounds simple? … but this is not always the case. Payment terms can also include the following.

- Dates that the Contractor invoices the employer. This could be a simple monthly invoice or could be related to the stages of a contract – that is, billing is on a percentage completion basis. Often a payment schedule is included in the contract. Stage payments could be linked to a programme of works (e.g. preliminary works, site works, laboratory testing, reporting).

- The Contractor may be required to issue an application for payment (AFP) which requires the approval of the employer prior to being able to submit an invoice. There may then follow a period before a 'due date for payment', issue of a 'payment notice' specifying an amount the employer considers due to the Contractor (subject to any pay less notice) and then a 'final date for payment' of the invoice.

- The employer may require alternative payment terms that fit in with a wider project team. There may also be terms for the Contractor to pay their subcontractors, which should not be more onerous than the Contractor's payment terms.

- The Contractor may have rights to claim interest on late payment and the employer may wish to provide a basis for this payment (e.g. to a percentage above the Bank of England base rate).

- Often payments for works are governed by Part 2 of The Housing Grants, Construction and Regeneration Act 1996, as amended by Part 8 of The Local Democracy, Economic Development and Construction Act 2009 (the Construction Act).

Cash flow is critical to the smooth running and financial reporting of a contract. Payment terms of 30 to 45 days from the date of invoice are typical. It should be noted that in some circumstances payments may be made in 'advance' rather than in 'arrears' where there may be a requirement for the early deployment of specialist equipment or the large cost of a particular item may affect a Contractor's cashflow.

Appendix C (continued on next page)

Contract clauses and provisions	Definitions and statements of use	Comments
		A pay less notice, as noted above, is a clause whereby the employer can inform the Contractor of any intention to pay less a certain period before the final date for payment. This should be a reasonable period of at least 5 days to give both parties the time to address any related issues before the proposed final date for payment. It is important to note that if the parties cannot agree on this then the Construction Act prevails, making it 7 days.
		Project bank accounts (PBA) are generally accepted as an official way to ensure prompt payment of Contractors. These are usually ring-fenced bank accounts, the sole purpose of which is to act as a channel for payment on construction projects to ensure that Contractors and other key members of the supply chain are paid on the contractually agreed dates. They are generally used in higher-value contracts where the parties enter a deed to join the PBA.
Bonds and parent company guarantees	RICS in their publication *Construction Security and Performance Documents* (RICS, 2013) note that bonds are forms of security that are often used in the construction industry. Numerous descriptions are given to them, such as 'on demand bond', 'conditional bond', 'default bond', 'guarantee', 'performance bond', 'retention bond' and 'parent company guarantee'. The most commonly used bonds in ground investigations are parent company guarantees and (rarer) performance bonds.	**Performance bond** can provide a Client with additional security against the Contractor's failure to perform. The Association of British Insurers provides a model form of guarantee bond for use in the UK construction industry. In addition, the bond amount required must be stated in the contract and is normally expressed as a proportion (typically 10%) of the contract value. The bond amount should be replicated in the bond itself, but Clients should be aware this is the maximum recoverable and not a guarantee that the full amount will be paid. Within the contract, Clients will need to decide the form of the bond and then include it in the scope. **Retention bond** issued under option (X16) of the NEC terms and conditions allows for retention to be released while still improving the cashflow for the Contractor and providing the Client with security for performance. These are expensive and rarely applicable to ground investigation contracts. There are also other forms of bond that may be included in the contract, such as offsite materials bonds. Most of these will not be relevant to ground investigation. The provision of bonds can be expensive for smaller Contractors and subcontractors and are generally not recommended for all but the largest of ground investigation contracts.

Parent company guarantee (PGC) may be a requirement in a particular contract clause and is usually in the form of a separate contract directly with the employer and the parent entity. PCGs are requested in order to gain financial or performance backing from the 'parent' – that is, the owner of the entity performing the work to stand behind and guarantee their ability to satisfy the Contractor's obligations, whether financial or otherwise. In the event of the subsidiary failing to perform its obligations under its contract with the beneficiary, the parent company may be required to perform its obligations in its place or reimburse the beneficiary for losses suffered because of the subsidiary's failure to perform.

Parent company guarantees (PGCs) create a financial obligation for an entity that is not otherwise a party to the transaction. They usually contain indemnities and other strict requirements. Some PCGs reference the 'parent company' while others may reference an 'ultimate parent company'. The parent would usually be the direct company above the contracting party (the owner of the business) whereby the ultimate parent company would usually be the highest level of ownership.

Often PCGs are required from subordinate smaller companies or companies without a good credit rating. Some specialist ground investigation firms may argue that they are a substantial business, with a high reported level of turnover, holding insurances commensurate to the services they provide and a PCG is unnecessary.

A PCG is a contract and, therefore, must contain all the necessary components of a contract, namely offer, acceptance, consideration and intent to create legal relations. The company/organisation giving the PCG should have the requisite capacity to offer the guarantee or else the provisions of the contract will not come into effect and the guarantee will be void and unenforceable.

Given a guarantor's liability under a guarantee is contingent on that underlying contract, any changes to that underlying contract can have the effect of releasing the guarantor from their obligations. This is particularly important in construction/engineering contracts, in which variations and instructions under the building contract are commonplace. Make the intent of the parties clear in the PCG by ensuring it expressly allows for amendments, variations and instructions under the building contract without affecting the enforceability of the guarantee.

Appendix C (continued on next page)

Contract clauses and provisions	Definitions and statements of use	Comments
Collateral warranties, letters of reliance, assignments and third-party agreements	There are often formal agreements that run alongside the main contract and create a contractual relationship with a third party that would otherwise not exist. Essentially the agreement gives a third party rights akin to those of the employer and increases the Contractor's liability.	It is important for the contractor to be given a reasonable amount of time to assess the conditions of a warranty/reliance letter/third-party agreement; legal advice or confirmation of the terms with an insurance broker is generally recommended. These documents associated with the increased business risks have a burden of administration in providing and keeping track of the ongoing responsibilities. The cost of the provision of these documents should be included in the bill of quantities. It should also be clear in the contract whether the terms of the document can be passed on to other parties with or without written consent; the reliance period (usually 6 or 12 years) should match that of the contract and the contractor's insurance coverage.
		Only the parties to a contract can generally sue to enforce the rights and obligations under it. This means that parties do not have to worry about unexpected claims from third parties. A collateral warranty (CW) or letter of reliance (LoR) are contracts under which the party undertaking services or works warrants to a third party that it has fulfilled its obligations under its professional appointment, contract or subcontract. The purpose of a collateral warranty is to give a third party, who is not a party to the original contract, rights to enforce that original contract. Without a contract an affected party may not be able to recover its loss due to the doctrine of privity of contract. It is important to note that the CW and LoR must not provide a greater or longer liability than that under the original appointment.
		The Joint Contracts Tribunal (JCT) produce several templates for setting up collateral warranties between various parties (these include a contractor/subcontractor collateral warranty for a funder, contractor/subcontractor collateral warranty for a purchaser or tenant and a subcontractor collateral warranty for the employer). These can be suitable documents for establishing reasonable terms for a collateral warranty within a ground investigation contract.

The introduction of the Contracts (Rights of Third Parties) Act 1999 was primarily aimed at the construction industry to dispense with the need for collateral warranties on each construction project. In broad terms, the Act provides that a person who is not a party to the contract may enforce a term of the contract in two situations

(a) if that contract expressly gives them the right to do so (with no further requirement of the relevant provision having been made for the benefit of the third party); or

(b) if the contractual term purports to confer a benefit on the third party.

There is a risk that the contractor/subcontractor will under the contract assume greater liabilities to third parties than they would have provided under collateral warranties, and it is imperative that the contracting party check the contract to identify the risks/liabilities being assumed under the contract to third parties. The Act has not had as big an impact as originally intended and is often expressly excluded in contracts using terms like the following: 'Notwithstanding any other provision of this contract nothing in this contract confers or purports to confer any right to enforce any of its terms on any person who is not a party to it.' There is still little case law on the operation of third-party rights and collateral warranties/reliance letters remain the most common mechanisms to give third parties contractual rights of recovery.

Assignment of the benefit of a ground investigation report, or the data contained in it, is often included as part of a CW or LoR. There can be other ways that the benefit is passed to other parties, and it should be clearly stated in the contract whether this is restricted or not. The AGS has published Loss Prevention Alert 11 *Confusion About Assignment* (AGS, 2000).

Third party agreements that the Client has entered can often be included within the contractual terms. These may be with landowners, statutory bodies or funders. It is important that the contractor's liabilities under these agreements are clearly set out within the contract to enable the contractor to effectively satisfy the terms.

Appendix C (continued on next page)

Contract clauses and provisions	Definitions and statements of use	Comments
Duty of care/fitness for purpose clauses	Where the standard of care in the provision of works/services or design responsibility is not clear in the specification or scope, the two common alternatives are variants of either (a) using 'reasonable skill and care' or (b) ensuring 'fitness for purpose'. An obligation to use 'reasonable skill and care' will generally be satisfied if the services provided or the design is prepared to a standard that a reasonably competent member of the profession could be expected to have achieved. In contrast, 'fitness for purpose' is an absolute commitment to achieve a result.	**Fitness for purpose**: A fitness for purpose obligation imposes a higher duty, as it is an absolute obligation to achieve a specified result, a breach of which does not require proof of negligence. The professional is effectively guaranteeing that the outcome of their work will be fit for its intended purpose. Fitness for purpose clauses can, therefore, operate in a seemingly harsh manner which results in the imposition of liability on a specialist in circumstances where they have acted with reasonable skill and care. Accepting this higher standard may not be achievable and may have insurance implications for the Contractor.

The contractual requirement should usually be that the standard for the ground investigation works is clearly identified in the specification or to use reasonable skill and care. It is not unusual that Contractors avoid accepting a 'fitness for purpose' obligation. Professional liability insurance typically only covers duties of 'reasonable skill and care', and most commercial parties are unwilling to carry uninsured risk. 'Reasonable skill and care' is, therefore, the generally accepted standard in most sectors.

Deleterious materials clauses: These often include an obligation to ensure that no deleterious materials are used in the project or in the works. These are only appropriate to those parties specifying materials for a ground investigation. Association of Geotechnical and Geoenvironmental Specialists Loss Prevention Alert 42 *Deleterious Materials Clauses* (AGS, 2010) provides guidance on these clauses.

Time is of the essence: Stating that time is of the essence in a commercial contract requires that relevant deadlines are strictly complied with. The party relying on that clause can terminate the agreement and, if appropriate, claim damages if the other party fails to perform the obligation in accordance with the date or time specified in the agreement. It is rare that this strict liability is relevant to a ground investigation contract where liquidated damages and programme requirements adequately deal with timescales. One example of where time is of the essence would be where welfare facilities are to be delivered to a site to provide an essential service for a single day's work. Time is of the essence as if these arrive the day after the work is undertaken, they are essentially useless. |

Retention	Retention refers to monies with held for a fixed period of time at the end of a contract. This can be an acceptable methodology to ensure the adherence to the specified works.	Retention of payments as a percentage of the money owed to the Contractor is a common occurrence in construction and ground investigation contracts. Retention should be fully justified within the context of a ground investigation to avoid penalising a Contractor who may have a significant financial responsibility to pay subcontractors. The retention percentage should be commensurate with the required delivery of information and the retention period should not be too long. Typically a retention of 5% with 'full retention released on submission of Contractor's final report' or 50% released on completion of fieldworks and 50% on submission of final report would be acceptable. A small percentage of less than 2.5% may be retained for completion of monitoring works.
Intellectual property/copyright clauses	Intellectual property (IP) refers to creative work which can be treated as an asset or physical property. Intellectual property rights fall principally into four main areas: copyright, trademarks, design rights and patents.	Intellectual property is an asset for any business. A business brand and the way they do things are important components of every business. Some forms of IP can be listed on the company's balance sheet; failure to adequately protect IP can have consequences for the bottom line.

Contracts frequently include provisions that require provision of an indemnity against any third-party claims for infringement of intellectual property by the author of copyright material and/or that the owner of the moral rights will not seek to enforce their moral rights against any other third party.

Contractors will often resist the assignment of the rights of ownership of intellectual property or the provision of an indemnity for copyright moral rights. This is often due to the imposed burden on the Contractor associated with the prevention of infringing the copyright. A more agreeable route to allow documents to be used for the purpose they were intended is to provide a clause that grants an irrevocable, fully transferable, royalty-free, nonexclusive licence, including a right to grant sublicences on equivalent terms, to copy and use the ground investigation documents. |

Appendix C (continued on next page)

Contract clauses and provisions	Definitions and statements of use	Comments
Set-off clauses	Set-off clauses are clauses in contracts that attempt to set off sums alleged to be due under one contract against a different contract.	It is important to distinguish that in terms of set-off, the position is different in construction contracts covered by the Construction Act than it might be in a general commercial contract for services. It is fair to assume that the Act will apply for ground investigation contracts. S110(1A) of the Act prohibits making payment conditional on the performance of obligations under another contract. Opinions are divided on the point, with some arguing that the clause does not expressly reference cross-contract set-off and does not cover the effect on other types of (noncontractual) set-off (such as set-off in equity), while others argue for the more literal reading of the words which prohibit such set-off.
		There are other avenues for set-off relating to construction contracts which are categorically decided, the most obvious being that parties cannot set off against an adjudicator's decision.
		It is important to note that there has been case law around the use of equitable set-off in similar scenarios, with the courts finding that a claim to set off sums from one contract against another failed unless the contracts were so closely connected with the claim that it would be 'manifestly unjust' to allow the claim without considering the cross-claim. Therefore, there may be some scenarios where there are two contracts so closely related to each other that an equitable set-off would be permitted. Project-specific set-off may be appropriate within a ground investigation, whereas global set-off is unlikely to be either fair or enforceable.

Temporary works

Temporary works are the elements of the construction that are needed to facilitate the construction of the permanent works and will, most often, be removed once construction is complete.

In ground investigation, a ground investigation Contractor's design responsibility is often limited to temporary works. The contract should detail the Contractor's liabilities for design and the associated insurance requirements, particularly on the rare occasion where the design is to be incorporated into the permanent works. If the design element becomes significantly larger than the site works, a professional service contract may be appropriate.

There may be aspects of future reliance on the Contractor's design that require some form of agreement to another party, which can be incorporated into collateral warranties, letters of reliance, assignments and third-party agreements, as discussed above.

Termination clauses

When both parties have performed all of their obligations under a contract, including all express and implied terms, a contract comes to an end.

Terms for termination prior to completion can include bankruptcy/liquidation, unrecoverable default of one of the parties to the contract, agreement of both parties to terminate, failure to make payments, unforeseen events and various other legal reasons (including breach of contract and frustration). It is important to clearly identify the reasons and process for termination so that it is fair and equitable (some NEC contracts have a 'Termination Table' setting out the reasons for termination, procedure and amounts due). Termination for the convenience of the Client, if included, would mean the contractor would not be entitled to any loss of profits due to the termination of the contract.

It is important that the data that has been obtained during a ground investigation contract prior to termination is protected and can still be used by the Client following the termination procedure.

Defects clauses

These allow for any defects in the work undertaken to be rectified during the period of the contract.

Defects liability (or correction) periods are frequently included within contracts to ensure that any defective work is repaired, replaced or made good prior to completion. It is generally accepted that a period of 6–12 months is necessary to confirm acceptable reinstatement and the correct functioning of installed equipment within a ground investigation. It is not usually required to retain significant payments from the Contractor as remediation is often a straightforward process.

Appendix C (continued on next page)

Contract clauses and provisions	Definitions and statements of use	Comments
'Back-to-back' provisions	Clients often require 'back-to-back' provisions in a contract (effectively mirroring and passing on their own contractual liabilities to the Contractor).	This can be effectively achieved if the risks are clearly described in the contract documents, enabling the Contractor to effectively price them. However, the party preparing the contract should be careful to avoid ambiguity for matters that could have a negative effect on the outcome of the project as a whole, such as the responsibility for design of the investigation, obtaining access to work areas, permit acquisition and liaising with landowners.

A term that states that a subcontractor has 'full knowledge' of a 'main contract' may satisfy a main Contractor's requirement to pass down all the contractual terms, but it is unlikely to be beneficial to achieving the best results, particularly if there are some onerous terms that are not immediately clear to the subcontractor. Designating that a copy of the main contract is available for the Contractor to read at a location a significant distance away from their premises is unnecessarily restrictive. It can hardly be considered as acting in a spirit of mutual trust and cooperation between the parties. |
| **Executing contracts as deeds** | A deed is a formal written contract. Deeds must be formally delivered, are valid even if not dated, can be enforced even though there is no consideration (payment) and often will not take effect until certain conditions are fulfilled (the deed is then referred to as a 'deed in escrow'). | Deeds are rarely necessary for ground investigation contracts as often the real reason for Clients wanting to execute contracts as deeds is to take advantage of a 12-year limitation period; after that amount of time the efficacy of a ground investigation may be limited. The Association of Geotechnical and Geoenvironmental Specialist have produced guidance on *The Dangers of Executing Contracts as Deeds* (AGS, 2007). |

Sustainability and climate change

Sustainability and climate change requirements are becoming increasingly important to governments and Clients. These may include carbon reduction policies organised around mitigation, adaptation and means of implementation and the four pillars of sustainability: human, social, economic and environmental.

There are a significant number of potential sustainability clauses that could be included in ground investigation contracts and these requirements are likely to grow in number as the stated time commitments of the key players draw nearer.

With a view to ensuring that climate change requirements are reflected in contracts and with a view to driving construction professionals to focus on sustainability and climate change, the NEC introduced a new Secondary Option clause for climate change, which has been summarised in an article in the AGS magazine (Beale & Co., 2024). Option X29 (climate change) has been drafted as a Secondary Option that parties using the NEC4 suite of contracts can choose to include in their contracts. Published in August 2022, Option X29 represents a very positive step in helping the construction industry help tackle climate change.

Option X29 has been drafted in a deliberately flexible manner that enables the parties to decide on the approach that they want to take on a project-by-project basis, allowing project-specific amendments to be made to standard positions. Therefore, the drafting (and implications) of Option X29 will need to be carefully considered on a project-by-project basis.

Beale & Co. (2024) highlight a few key points to be considered in relation to Option X29 before it is used.

■ The clause is not a complete solution that can just be inserted into a contract. It requires a great deal of thought and discussion between the parties as to how it will be used and the targets that will be set. These targets need to be both attainable and measurable for the clause to be used effectively.

■ The definitions of 'climate change requirements' and the 'performance table' do not include a reference to any of the emerging standards or definitions that are being developed, such as the UK or World Green Building Council's definition of 'net zero'. While the drafting promotes flexibility, it does enable the Client to specify the determination of each definition. Contractors and consultants should assess the definitions and metrics for each and every contract.

■ Consultants and contractors will need to carefully consider the requirements of the clause in line with their professional indemnity insurance policies. There is a risk that some of the provisions may be inconsistent with insurance cover.

Appendix C Continued

Contract clauses and provisions	Definitions and statements of use	Comments
Additional clauses	There are potentially an unlimited number of additional clauses that can be added to a ground investigation contract.	Some of the typical topics are as follows ■ antislavery and human trafficking requirements ■ counter-terrorism and security ■ equality and diversity ■ nondisclosure (of confidential information), General Data Protection Regulations (GDPR) and document destruction including the Information Act 2000 (FOIA) and the Environmental Information Regulations 2004 (EIRs) ■ regulatory licenses ■ special requirements in relation to statutory bodies and others (service providers, Environment Agency (EA), highways, railways etc.) ■ additional conditions for work over water/Coal Authority/NRSWA/rail/, waste management requirements, Environment Agency Floodplain Permits, Japanese knotweed and so on.

AGS (Association of Geotechnical and Geoenvironmental Specialists) (2000) *Loss Prevention Alert 11. Confusion About Assignment.*

AGS (2007) *Loss Prevention Alert 37. Dangers of Executing Contracts as Deeds.*

AGS (2010) *Loss Prevention Alert 42. Deleterious Materials Clauses.*

AGS (2023) *Loss Prevention Alert 77. Limits of Liability – Are They 'Reasonable' Under the Unfair Contract Terms Act?*

Beale & Co (2024) *NEC Option X29 – A positive step to tackling climate change, but not without its risks.* AGS, Faversham, UK. https://www.ags.org.uk/2024/05/nec-option-x29-a-positive-step-to-tackling-climate-change-but-not-without-its-risks/

Royal Institution of Chartered Surveyors (RICS) (2013) *Construction security and performance documents.* RICS, London.

Association of Geotechnical and Geoenvironmental Specialists
ISBN 978-1-83662-029-7
https://doi.org/10.1108/978-1-83662-026-620251010

Appendix D
Glossary of ICC terms

Appendix D (continued on next page)

Term	Definition
Acceptance Certificate	As defined in Clause 61(1).
Ancillary Works	All appliances or things of whatsoever nature required to be installed or constructed on under in or through the Site and which are to remain on Site and become the property of the Employer in accordance with the Contract.
Bill of Quantities	The priced and completed itemised list of the quantity of works required to be carried out by the Contractor to complete the contract.
Certificate of Substantial Completion	A certificate issued under Clause 48.
Commencement Date	As defined in Clause 41(1).
Contract	The Conditions of Contract, Specification Schedules, Drawings, Bill of Quantities, the Form of Tender, the written acceptance thereof and the Form of Agreement (if completed).
Contract Price	The sum to be ascertained and paid in accordance with the provisions for carrying out the Investigation in accordance with the Contract.
Contractor	The person or persons, firm or company to whom the Contract has been awarded by the Employer and includes the Contractor's personal representatives, successors and permitted assignees.
Contractor's Equipment	All appliances or things of whatsoever nature required in or about the carrying out of the Site Operations but does not include materials or other things intended to form or forming part of the Ancillary Works.
Defects Correction Period	The period stated in the Appendix to the Form of Tender calculated from the date on which the Contractor becomes entitled to a Certificate of Substantial Completion for the Site Operations or any Section or part thereof.
Drawings	The drawings referred to in the Specification and any modification of such drawings approved in writing by the Engineer and such other drawings as may from time to time be furnished by or approved in writing by the Engineer.
Employer	The person or persons, firm, company or other body named in the Appendix to the Form of Tender and includes the Employer's personal representatives, successors and permitted assignees.
Engineer	The person, firm or company appointed by the Employer to act as Engineer for the purposes of the Contract and named in the Appendix to the Form of Tender or any other person, firm or company so appointed from time to time by the Employer and notified in writing as such to the Contractor.
Engineer's Representative	A person notified as such from time to time by the Engineer under Clause 2(3)(a).

Ground Specialist
A person qualified as a Chartered Engineer or Geologist or Environmental Scientist with at least five years of relevant experience.

Investigation
The Site Operations together with the Laboratory Testing and Report preparation and submission.

Laboratory Testing
The testing operations and processes necessary for the preparation of the Report to be carried out in accordance with the Contract at a laboratory approved by the Engineer on samples and cores obtained during the Site Operations.

Nominated Contractor
Any merchant, tradesman, specialist or other person, firm or company nominated in accordance with the Contract to be employed by the Contractor for the carrying out of work or supply of goods, materials or services for which a Prime Cost or a Provisional Sum has been included in the Contract.

Prime Cost (PC) Item
An item in the Contract which contains (either wholly or in part) a sum referred to as Prime Cost (PC) which will be used for the carrying out of work or the supply of goods, materials or services for the Investigation.

Provisional Sum
A sum included and so designated in the Contract as a specific contingency for the carrying out of work or the supply of goods, materials or services which may be used in whole or in part or not at all at the direction and discretion of the Engineer.

Report
The report to be prepared and submitted in accordance with the Contract.

Schedules
The schedules and lists of Site Operations, Laboratory Testing and other requirements referred to in the Specification.

Section
A part of the Investigation separately identified in the Appendix to the Form of Tender.

Specification
The specification referred to in the Form of Tender and any modification thereof or addition thereto as may from time to time be furnished or approved in writing by the Engineer.

Site
The lands and other places on, under, in or through which the Site Operations are to be carried out and any other lands, places or access provided by the Employer for the purposes of the Contract together with such other places as may be designated in the Contract or subsequently agreed by the Engineer as forming part of the Site.

Site Operations
All work of every kind including Ancillary Works required to be carried out on, under, in or through the Site in accordance with the Contract.

Tender Total
The total of the Bill of Quantities at the date of award of the Contract or in the absence of a Bill of Quantities the agreed estimated total value of the Investigation at that date.

Association of Geotechnical and Geoenvironmental Specialists
ISBN 978-1-83662-029-7
https://doi.org/10.1108/978-1-83662-026-620251011
Emerald Publishing Limited: All rights reserved

Appendix E
Glossary of NEC terms

Courtesy of NEC Dictionary of Terms https://www.neccontract.com/resources/dictionary

Accepted Programme	The Accepted Programme (in the ECC) is identified within the Contract Data or during the project it will be superseded by a later programme – there is a process for submission and this must be accepted by the project manager. The programme is to be practicable and realistic, showing when the Contractor intends to carry out each part of the works, identifying the resource it intends to use – this tool is invaluable in successfully managing a contract.
Compensation Events	NEC3 and NEC4 contracts are underpinned by a philosophy that the Contractor (in the ECC) should not lose out, nor benefit from a windfall, because of an event which occurs and is at the Client's risk. Compensation events entitle a Contractor to be compensated for any impact the event has on the prices, completion or key dates in the contract. The cost impact of a compensation event is based on the effect the event has on the cost of work already done and the forecast cost of work not yet done at the dividing date, which is often when the project manager instructed the change to the scope.
Completion	This is when all work that the scope (in the ECC) states is to be completed by the Completion Date, including correcting any notified Defects.
Contract	NEC3 and NEC4 can be used during the entire project life cycle; from planning, defining legal relationships and procurement of works, all the way through to project completion, FM or asset maintenance and beyond. The intended outcome of NEC3 contracts is that time and money performances are improved while increasing standards by encouraging collaborative working in order to achieve shared project objectives. This philosophy promotes a less adversarial approach, decreasing the chance and impact of costly disputes.
Contract Data	This information should be completed as part of the necessary contract documents; in the ECC this contains details such as starting date/Completion Date/employer details/defects date and the like. It is in two parts; part one (provided by the Employer) and part two (provided by the Contractor).

Core Clauses	Most NEC4 contracts have nine core clauses which follow the list below. 1. General terms 2. The Contractor's main responsibilities 3. Time 4. Quality management 5. Payment 6. Compensation events 7. Title 8. Liability and insurance 9. Termination.
Defect	In the ECC, this is a part of the works which is not as stated in the scope or not in accordance with applicable law or the accepted design. There is a reciprocal obligation on both the Supervisor and Contractor to notify each other as soon as they are aware of a defect. At an agreed date, the project Supervisor will list any uncorrected defects or certify that there are no defects (defects certificate).
Defects Certificate	In the ECC, the Supervisor issues this at or just after the defects date. The certificate is a statement that there are uncorrected defects or that there are no defects; this starts the closing down of the contract.
Defined Cost	A defined term used in the ECC and other NEC3 or NEC4 contracts to give the parties a definition of Contractor's cost for different use in different main Options. They only include the amounts calculated using rates and percentages in the Contract Data and other amounts at open market/ tendered /discounted rates.
ECC	Engineering and Construction Contract.
ECS	Engineering and Construction Subcontract.
Early Warning	The Contractor and project manager should notify each other, as soon as practicable, of any matter which could affect the cost, completion, progress or quality of the works (in the ECC).
Main Options	These are the pricing mechanisms that the main NEC3 and NEC4 contracts offer to choose from a tender stage. These Options include lump sum, target cost or cost reimbursable contracts.
NEC3	NEC3 and NEC4 contracts are a family of standard contracts, each of which stimulates good management, are flexible and are clear and simple.
NEC4 **Option A**	Option A: priced contract with activity schedule, which is used with the ECC, ECS, TSC and PSC (for NEC3) and the ECC, ECS, TSC and PSC (for NEC4).
Option B	Option B: priced contract with bill of quantities used with the ECC and ECS for NEC3 and with the ECC and ECS for NEC4.
Option E	Option E: cost-reimbursable contract used with the ECC, ECS and TSC for NEC3 and with the ECC, ECS, TSC and PSC for NEC4.
Project Manager	Has a role in managing time, testing and defects, payments, compensation events, title, risks and insurance and finally termination of a contract if it is considered. However, it is the Employer and Contractor who are in contract. The project manager 'administers' the contract.

PSC	Professional Service Contract.
Scope	A defined term in the ECC, this includes the specification and description of works the Contractor is to provide; it also might include a series of constraints to which the Contractor must adhere to. It will be included either in a place specified by the contract data or later amended by a project manager's instruction.
Supervisor	A named role in the ECC, the Supervisor is responsible for notifying Defects to the Contractor, witnessing tests and inspections by the Contractor and monitoring the quality of construction works on site.
TSC	Term Service Contract.
Working Areas	A defined term in the ECC to describe those areas which are necessary for, and only used to deliver, the required works.
Works	Various NEC3 and NEC4 contracts deal with the provision of works. This encompasses contracts for the construction, refurbishment and decommissioning of buildings, structures, process plants and infrastructure – including everything from houses, schools, hospitals and leisure facilities to infrastructure for water, energy, transport, industry and waste.
Z Clause	This is a Secondary Option available in several NEC3 and NEC4 contracts. These allow for additional conditions of contract to that in the printed form – this could comprise of additions, deletions or omissions. Careful consideration should be given to the need for a Z clause. Care should also be taken to ensure that the clause retains the clarity, style and terminology of the rest of the contract as well as that it does not cause conflict with other clauses.

emerald PUBLISHING   ice

Association of Geotechnical and Geoenvironmental Specialists
ISBN 978-1-83662-029-7
https://doi.org/10.1108/978-1-83662-026-620251012

Appendix F
Potential areas of conflict for ground investigations

Potential conflict/issue	Planning strategy	Outcome
Misunderstanding or miscommunication by using complicated language/terms	Keep language simple and relevant.	Allows for clearer instructions and requirements. Ambiguity reduced or eliminated.
Instructions received verbally	Provide written instructions from outset. Follow up any conversation/meeting with written minutes or instructions.	A clear record of instructions.
Unrealistic timescales	Create an agreeable timescale to complete all activities.	Increased appreciation of realistic timescales.
Unclear reporting or management lines	Provide a clear operational organogram. Agree to core teams/liaison between each party.	Identification of key individuals and their responsibilities.
Insufficient time allowed for planning	Programme sufficient time and adopt ECI stage.	Reduces or eliminates potential delays.
Misunderstanding each party's 'values'	Document the values and align where possible.	All parties aware of each other's values.
Differing opinions/experience	Understand the objectives and adopt collaborative working to mutual end goal.	Integrating each party's experience and skills to meet the objectives.
Ambiguous responsibilities	Define each party's responsibility (contractual and operational).	Each party and their representatives understand their documented role.
Unclear communication	Define responsible persons and processes for communication flow.	Clear and agreed lines of communication.
Programme deviation	Agree realistic timescales at the start. Provide early warnings if delays are anticipated. Organise regular meetings to discuss programme.	Management of expectations for adherence to programme.

Potential conflict/issue	Planning strategy	Outcome
Provision of inappropriate facilities for Client/Designer	Ensure these are allowed for and clearly defined in bill of quantities, not just in the specification.	Appropriate facilities provided for the site and operations.
Unforeseen ground conditions	Inform all parties of any unexpected change in geology/ground conditions. Keep good records/take photos. Discuss at regular meetings.	All parties made aware, as soon as practicable.
Change in scope	Discuss and mutually agree any change in scope at pertinent/suitable time.	Change in scope managed on an ongoing basis.
Standing time	Define and agree to clear terms for standing of plant and personnel. Record and make parties aware when standing occurs or is likely to occur.	Standing time agreed as works progress or avoided if alternative solutions are possible.
Reinstatement	Pre-/post-condition surveys to be defined within the specification and sufficient time/ costs allowed for.	All parties aware of reinstatement requirements and clear records provided.
Increase of costs	Transparent bill of quantities/ calculation of costs. Regular communications and, if necessary, meetings providing details of cost increases. Timely invoicing and approvals.	Provision in a clear format of all works completed.

Association of Geotechnical and Geoenvironmental Specialists
ISBN 978-1-83662-029-7
https://doi.org/10.1108/978-1-83662-026-620251013

Index